WHY DO CLOCKS RUN CLOCKWISE?

ALSO BY DAVID FELDMAN

Do Elephants Jump?

How Do Astronauts Scratch an Itch?

What Are Hyenas Laughing at, Anyway?

How to Win at Just About Anything

How Does Aspirin Find a Headache?

Are Lobsters Ambidextrous?

  (Originally published as *When Did Wild Poodles Roam the Earth?*)

Do Penguins Have Knees?

Why Do Dogs Have Wet Noses?

When Do Fish Sleep?

Who Put the Butter in Butterfly?

Why Don't Cats Like to Swim?

  (Originally published as *Imponderables*)

WHY DO CLOCKS RUN CLOCKWISE?
An Imponderables® Book

David Feldman

Illustrated by Kassie Schwan

Perennial Currents
An Imprint of HarperCollins*Publishers*

For Michael Leo Feldman

"Imponderables" is a trademark of David Feldman.

A hardcover edition of this book was published in 1987 by Harper & Row, Publishers.

First Perennial Currents edition published 2005.

Designer: Sidney Feinberg
Copyeditor: Abigail Bok

Library of Congress Cataloging-in-Publication Data
Feldman, David.
    Why do clocks run clockwise? and other imponderables.
    Includes index.
    1. Questions and answers. I. Title.
    AG195.F46 1988   031'02   87-45045
    ISBN 0-06-091515-3
    ISBN 0-06-074092-2 (pbk.)

05 06 07 08 09 RRD 10 9 8 7 6 5 4

Contents

CONTENTS

CONTENTS   ix

doughnuts. Why do they then put the same tissue in with the doughnuts for the customer to carry home—germs and all?

Preface

If you read the first volume of *Imponderables*, you now know why you don't ever see baby pigeons, why women open their mouths while applying mascara, and why people look up when thinking. But the last frontiers in human knowledge haven't quite yet been plumbed. Thus the burning need for *Why Do Clocks Run Clockwise? and Other Imponderables*.

Imponderables are the everyday mysteries of life that aren't very important—until they occur to you. Then they begin to gnaw at your brain like termites boring through wood. An Imponderable is a mystery that cannot be solved by numbers or measurements or standard reference books. You will sleep better when you find out where flies go in winter, what happens to the tread that wears off tires, and why hamburger-bun bottoms are so thin that they disintegrate under the weight of the patty. You will be a better person for knowing this stuff.

Most of the new Imponderables come from readers of the first book. It is humbling to discover that the readers of *Imponderables* are at least as imaginative (crazed?) as its author.

Many readers asked for a subject index in the next volume of *Imponderables*. Your wish is our command. We've also added a new feature, "Frustables," short for Frustrating Imponderables, a top-ten list of Imponderables whose answers eluded us. We are offering a free copy of the next edition of *Imponderables* to the first person who provides evidence or referrals that lead us to solutions of these ultimate mysteries.

And, of course, we still offer a free copy of the next edition of *Imponderables* to the first person who poses an Imponderable we answer in our next volume. The last page of the book will tell you how you can unburden your soul of the mysteries that plague you and participate in this great intellectual journey. But for now, sit back and enjoy.

What Is the Purpose of the Warning Label on a Mattress? And What Happens If I Rip It Off?

Here is an Imponderable that happens to be one of the foremost moral issues plaguing our society today. Many transgressors are consumed with guilt over having ripped off mattress tags. Some are almost as upset about impetuously doing in pillow tags, as well.

We are here to say: do not be hard on yourself. You have done nothing legally wrong. You have not even done anything morally wrong.

Those warning labels are there to protect you, not to shackle you. If you look carefully at the language of the dire warning, there is always a proviso that the label is not to be removed "except by the consumer." Labeling laws are up to the individual states. Thirty-two of the fifty states have laws requiring mattress tags, and none of the states cares whether the purchaser of a mattress rips up the tag.

So how do these warning labels protect you? Most important, they inform the consumer exactly what the filling material is made of, because the fill is not visible. The label also notifies the consumer that the manufacturer is registered with all of the appropriate government agencies and has fulfilled its obligations in complying with their regulations. There is also manufacturing information on the tag that may help the consumer when and if a warranty adjustment is desired (though this is a good argument for keeping the tag on the mattress, or at least filing it for future reference).

One of the reasons why mattress warning label laws were imposed in the first place is that some less-than-ethical merchants used to palm off secondhand mattresses as new ones. It is legal, in most states, to sell secondhand mattresses as long as they are properly sterilized. A white tag guarantees a new mattress; a sterilized secondhand mattress carries a yellow tag.

Submitted by the Reverend Ken Vogler, of Jeffersonville, Indiana. Thanks also to: Mike Dant, of Bardstown, Kentucky, and Owen Spann, of New York, New York.

Why Do Dogs Walk Around in Circles Before Lying Down?

The most common and logical explanation for the phenomenon is that in the wild, circling was a method of preparing a sleeping area or bed, particularly when it was necessary to flatten down an area among tall grass, leaves, and rocks.

Some experts also believe that circling is a way for dogs to map territory, to define an area of power. Dog writer Elizabeth Crosby Metz explains the habit this way:

I believe it also has to do with spreading their proprietary scent around their nesting site, to say: "Keep away, this is MY nest!"

In fact, as a breeder I know that mother dogs will circle many times before lying down to feed their sightless, deaf newborns as a way of spreading her scent and indicating to them exactly where she is and how far they have to go to reach her. Think about it: How else can blind, deaf newborns so surely find the milk bar?

Submitted by Daniel M. Keller, of Solana Beach, California. Thanks also to: Joanna Parker, of Miami, Florida.

If Nothing Sticks to Teflon, How Do They Get Teflon to Stick to the Pan?

"They," of course, is Du Pont, which owns the registered trademark for Teflon and its younger and now more popular cousin, Silverstone. G. A. Quinn, of Du Pont, told *Imponderables* that the application of both is similar:

> When applying Silverstone to a metal frypan, the interior of the pan is first grit-blasted, then a primer coat is sprayed on and baked. A second layer of Polytetrafluoroethylene (PTFE) is applied, baked and dried again. A third coat of PFTE is applied, baked and dried.
>
> About the only thing that sticks to PTFE is PTFE. So, the 3-coat process used in Silverstone forms an inseparable bond between the PTFE layers and the primer coat bonds to the rough, grit-blasted metal surface.

Du Pont has recently introduced Silverstone Supra, also a three-layer coating that is twice as durable as conventional Silverstone.

Submitted by Anthony Virga, of Yonkers, New York.

Why Is the Scoring System in Tennis So Weird?

Tennis as we know it today is barely over a hundred years old. A Welshman, Major Walter Clopton Wingfield, devised the game as a diversion for his guests to play on his lawn before the real purpose for the get-together—a pheasant shoot. Very quickly, however, the members of the Wimbledon Cricket Club adopted Wingfield's game for use on their own underutilized lawns, empty since croquet had waned in popularity in the late eighteenth century.

Long before Wingfield, however, there were other forms of tennis. The word "tennis" first appeared in a poem by John Gower in 1399, and Chaucer's characters spoke of playing "rackets" in 1380. Court tennis (also known as "real" tennis) dates back to the Middle Ages. That great athlete, Henry VIII, was a devotee of the game. Court tennis was an indoor game featuring an asymmetrical rectangular cement court with a sloping roof, a hard ball, a lopsided racket, and windows on the walls that came into play. Very much a gentleman's sport, the game is still played by a few diehards, though only a handful of courts currently exist in the United States.

Lawn tennis's strange scoring system was clearly borrowed from court tennis. Although court tennis used a fifteen-point system, the scoring system was a little different from modern scoring. Each point in a game was worth fifteen points (while modern tennis progresses 15–30–40–game, court tennis progressed 15–30–45–game). Instead of the current three or five sets of six games each, court tennis matches were six sets of four games each.

The most accepted theory for explaining the strange scoring system is that it reflected Europeans' preoccupation with astronomy, and particularly with the sextant (one-sixth of a circle). One-sixth of a circle is, of course, 60 degrees (the number of points in a game). Because the victor would have to win six sets of four games each, or 24 points, and each point was worth 15 points, the game concluded when the winner had "completed" a circle of 360 degrees (24 × 15).

Writings by Italian Antonio Scaino indicate that the sextant scoring system was firmly in place as early as 1555. When the score of a game is tied after six points in modern tennis, we call it "deuce"—the Italians already had an equivalent in the sixteenth century, *a due* (in other words, two points were needed to win).

Somewhere along the line, however, the geometric progression of individual game points was dropped. Instead of the third point scoring 45, it became worth 40. According to the *Official Encyclopedia of Tennis*, it was most likely dropped to the lower number for the ease of announcing scores out loud, because "forty" could not be confused with any other number. In the early 1700s, the court tennis set was extended to six games, obscuring the astronomical origins of the scoring system.

When lawn tennis began to surpass court tennis in popularity, there was a mad scramble to codify rules and scoring procedures. The first tennis body in this country, the U.S. National Lawn Tennis Association, first met in 1881 to establish national standards. Prior to the formation of the USNLTA, each tennis club selected its own scoring system. Many local tennis clubs simply credited a player with one point for each rally won. Silly concept. Luckily, the USNLTA stepped into the breach and immediately adopted the English scoring system, thus ensuring generations of confused and intimidated tennis spectators.

There have been many attempts to simplify the scoring system in order to entice new fans. The World Pro Championship League tried the table-tennis scoring system of twenty-one–point matches, but neither the scoring system nor the League survived.

Perhaps the most profound scoring change in this century has been the tie breaker. The U.S. Tennis Association's Middle States section, in 1968, experimented with sudden-death playoffs, which for the first time in modern tennis history allowed a player who won all of his regulation service games to lose a set. The professionals adopted the tie breaker in 1970, and it is used in almost every tournament today.

Submitted by Charles F. Myers, of Los Altos, California.

Why Have Humans Lost Most of Their Body Hair?

Anthropologists have debated this issue for a long time. Hair on most creatures is an important means of maintaining heat in the body, so the reasons for humans losing this valuable form of insulation are unlikely to be trivial. Here are some of the more logical theories advanced about why hair loss made sense for humans, along with the opinions of Desmond Morris, whose book *The Naked Ape* derives its title from this very Imponderable.

1. Hair loss allowed primitive people to cope better with the myriad of skin parasites, such as ticks, mites, and vermin, that bothered them. Parasites were more than a nuisance; they spread many potentially fatal infectious diseases. Although this theory makes sense, it doesn't explain why other relatives of man, equally bothered by parasites, have not evolved similarly.

2. Naked skin could have been a social rather than a func-

tional change. Most species have a few arbitrarily selected characteristics that differentiate them from other species—what Desmond Morris calls "recognition marks." Morris doubts the validity of the "recognition mark theory," for hair loss is a far more drastic step than is necessary to differentiate humans from other primates.

3. Hair loss might have had a sexual and reproductive basis. Male mammals generally are hairier than their female counterparts. This type of sex-based physiological difference helps make one sex more attractive to the other. Morris also mentions that hair loss served to heighten the excitement of sex—there is simply more tactile sensation without fur. Now that we are in the midst of a worldwide population explosion and trying to slow the birth rate, it is easy to forget that nature has built into our species characteristics to help increase our numbers.

4. Some anthropologists believe that before humans became hunting animals, stalking the savannas of East Africa, we went through a phase as an aquatic animal, seeking food at tropical seashores rather than on the more arid open plains. Without hairy bodies, humans became more streamlined in the water, able to swim and wade effectively. This is also a possible explanation for our hair being so plentiful on the head: if we spent much of our time wading in the water, only the top of the head need be covered, as protection against the sun. According to this theory, man left the water only after he developed the tools necessary to hunt.

5. Even if the aquatic phase never existed, hair loss helped humans regulate their body temperature after they moved from the forests to a plains-based hunting culture. Morris questions this theory: after all, other mammals, such as lions and jackals, made a similar switch of terrains without accompanying hair loss. Furthermore, the loss of body hair had a negative effect in that it subjected humans to dangerous ultraviolet radiation from the sun.

6. Hair loss kept primitive humans from overheating during the chase when hunting. This is Morris's pet theory. When our

ancestors became hunters, their level of physical activity increased enormously. By losing their heavy coat of hair, and by increasing the number of fat and sweat glands all over the body, humans could cool off faster and more efficiently. Sweat glands could not deliver their cooling effect nearly as effectively if fur trapped perspiration.

As divorced as we are from the problems of our primate forebears, Morris believes that human genetics will always countermand our attempts to elevate our culture. "[Man's] genes will lag behind, and he will be constantly reminded that, for all his environment-moulding achievements, he is still at heart a very naked ape."

Submitted by Sean S. Gayle, of Slidell, Louisiana.

Why Don't People Get Goosebumps on Their Faces?

Be proud of the fact that you don't get goosebumps on your face. It's one of the few things that separate you from chimpanzees.

We get goosebumps only on parts of our bodies that have hair. As you have learned from the exciting previous entry, the purpose of body hair is to protect us from the cold. But when our hair doesn't provide enough insulation, the small muscles at the bottom of each hair tighten, so that the hair stands up.

In animals covered with fur, the risen strands form a protective nest of hairs. Cold air is trapped in the hair instead of bouncing against delicate skin. The hair thus insulates the animals against the cold.

Although humans have lost most of their body hair, the same muscular contractions occur to defend against the cold. Instead of a mat of hair, all we have to face the elements are a few wispy tufts and a multitude of mounds of skin, which used to support an erect hair and now must go it alone. When a male lion gets "goosebumps," his erect hair makes him ferocious; our goose-bumps only make us look vulnerable.

Submitted by Pam Cicero, of Madison, Ohio.

Why Doesn't Countdown Leader on Films Count All the Way to One?

Remember watching the leader on sixteen-millimeter films in school, waiting for the countdown to go 10–9–8–7–6–5–4–3–2–whoops? It *never* got down to one.

Countdown leader, of course, is there to help the projectionist time when a film is going to start. Each number is timed to appear precisely one second after the other. The projectionist usually uses the number two as the cue to allow the projector light to hit the screen and begin the show. What would be number one is simply the start of the picture.

Wouldn't it work just as well to have zero represent the beginning of the movie, so that frustrated audiences could have the satisfaction of counting down from ten to one? Of course it would. But as in most areas, tradition and inertia rule. As Bob Dylan wrote, "Don't follow leaders."

Submitted by Ronald C. Semone, of Washington, D.C.

When a Company Sells Lobster Tails to Restaurants and Stores, What Do They Do with the Rest of the Lobster?

The American or Maine lobster is usually delivered to stores and restaurants whole and served in the shell. The claws, legs, and body all contain good lobster meat, and aficionados covet the green "tomalley" or liver and the red roe often found in female lobsters. As Richard B. Allen, vice-president of the Atlantic Offshore Fishermen's Association, put it, "If they are not eating almost everything except the shell, they are missing a fine eating experience."

If you find "lobster tails" listed on a restaurant menu, chances are you are ordering rock or spiny lobsters. Unlike Maine lobsters, spiny lobsters do not have big claws, but rather two large antennas. We spoke to Red Lobster's purchaser, Bob Joseph, who told us that most of Red Lobster's tails come from Brazil, Honduras, and the Bahamas. Red Lobster buys about two million pounds of spiny lobster a year (as well as a similar poundage of Maine lobster). The nontail meat of the spiny lobster is stringy and watery compared to the tails, and not as good-looking. Consumers seem to prefer the big, steaklike chunks of the tail rather than the shreds of meat found on other parts of the crustacean.

Although some restaurants and fish stores will buy the relatively small claws of spiny (and rock) lobsters, what happens to the nontail meat that isn't in demand? The claw, thorax, and head meat is sold as "meat packs," which are used for soups (lobster bisque, gumbos) and reconstituted for seafood salads. Seafood and Italian restaurants often use "meat-pack" lobster for pastas. And, surprisingly, one of the biggest users of lobster "meat pack" is egg-roll makers.

Submitted by Claudia Wiehl, of North Charleroi, Pennsylvania.

Why Do They Need Twenty Mikes at Press Conferences?

If you look carefully at a presidential press conference, you'll see two microphones. But at other press conferences, you may find many more. Why the difference?

Obviously, all the networks have access to the president's statements. How can they each obtain a tape when there are only a couple of microphones? They use a device called a "mult box" (short for "multiple outlet device"). The mult box contains one input jack but numerous output jacks (usually at least eight outputs, but sixteen- and thirty-two–output mult boxes are common). Each station or network simply plugs its recording equipment into an available output jack and makes its own copy. The second microphone is used only as a backup, in case the other malfunctions. The Signal Corps, which runs presidential news conferences, provides the mult boxes at the White House.

It's more likely, though, that a press conference will be arranged hastily or conducted at a site without sophisticated electronics equipment. It is at such occasions that you'll see multiple microphones, with each news team forced to install its own equipment if it wants its own tape.

All networks and most local television stations own mult boxes. Of course, the whole purpose of the mult box is to promote pooling of resources, so the networks, on the national level, and the local stations, in a particular market, alternate providing mult boxes. There usually isn't a formal arrangement for who will bring the mult box; in practice, there are few hassles.

Some media consultants like the look of scores of microphones, believing it makes the press conference seem important. A more savvy expert will usually ask for a mult box, so that the viewing audience won't be distracted by the blaring call letters on the microphones from a single pearl of wisdom uttered by the politician he works for.

Why Do Some Localities Use Salt and Others Use Sand to Treat Icy Roads?

Gabriel Daniel Fahrenheit, creator of the Fahrenheit temperature scale, discovered that salt mixed with ice (at a temperature slightly below the freezing point) creates a solution with a lower freezing point than water alone. Thus, salt causes snow and ice to melt.

Most localities haven't found a better way to de-ice roadways and sidewalks than salt. Salt is also effective in keeping hard packs of ice from forming in the first place. While a number of chemicals have been developed to melt ice, salt remains a much cheaper alternative.

So why don't all localities use salt to treat icy roads? Ecological problems have led some municipalities to ban the use of salt outright. Salt causes corrosion of vehicles, pavement, bridges, and any unprotected steel in surrounding structures. Salt also harms many kinds of vegetation.

The effectiveness of salt as an ice remover also has distinct limitations. Salt is best used in high-traffic areas; without enough traffic to stimulate a thorough mixing of ice and salt, hard packs can still develop. Below approximately 25° F, salt isn't too effective, because ice forms so fast that the salt doesn't have a chance to lower the freezing point. And salt applied on top of ice doesn't provide traction for drivers or pedestrians.

Sand, by contrast, provides excellent traction for vehicles when grit comes in contact with tires, whether the sand is exposed on top of the surface of the ice or mixed in with slush or snow. Sand doesn't require high-volume traffic areas to work effectively, it does little or no harm to vegetation, vehicles, or road, and is (pardon the expression) dirt cheap.

There is only one problem with sand: it doesn't melt snow or ice. Salt tries to cure the problem. Sand attempts to treat the symptoms.

Some localities have experimented with sand-salt combinations. Actually, most sand spread on pavements already contains some salt, used to keep the sand from freezing into clumps when mixed with the snow.

While salt is considerably more expensive than sand, cost is rarely the main criterion for choosing sand over salt. Joseph DiFabio, of the New York State Department of Transportation, told *Imponderables* that salt costs approximately twenty dollars per ton, compared to five dollars per ton for sand. In fact, sand must be applied in a greater concentration than salt, about three times as much. Because maintenance crews must use three times as much sand to treat the same mileage of roadway, they must return to reload their trucks with sand three times as often as with salt. The eventual cost differential, therefore, is negligible.

Submitted by Daniel A. Placko, Jr., of Chicago, Illinois.

Why Is the Telephone Touch-Tone Key Pad Arranged Differently from the Calculator Key Pad?

Conspiracy theories abound, but the explanation for this Imponderable reinforces one of the great tenets of Imponderability: when in doubt, almost any manmade phenomenon can be explained by tradition, inertia, or both. A theory we have often heard is that the phone company intentionally reversed the calculator configuration so that people who were already fast at operating calculators would slow down enough to allow the signals of the phone to register. It's a neat theory, but it isn't true. Even today, fast punchers can render a touch-tone phone worthless.

Both the touch-tone key pad and the all-transistor calculator were made available to the general public in the early 1960s. Calculators were arranged from the beginning so that the lowest digits were on the bottom. Telephone keypads put the 1–2–3 on the top row. Both configurations descended directly from earlier prototypes.

Before 1964, calculators were either mechanical or electronic devices with heavy tubes. The key pads on the first calculators actually resembled old cash registers, with the left row of keys numbering 9 on top down to 0 at the bottom. The next row to the right had 90 on top and 10 on the bottom, the next row to the right 900 on top, 100 on the bottom, and so on. All of the early calculators were ten rows high, and most were nine rows wide. From the beginning, hand-held calculators placed 7–8–9 on the top row, from left to right.

Before the touch-tone phone, of course, rotary dials were the rule. There is no doubt that the touch-tone key pad was designed to mimic the rotary dial, with the "1" on top and the 7–8–9 on the bottom. According to Bob Ford, of AT&T's Bell Laboratories, a second reason was that some phone-company research concluded that this configuration helped eliminate dialing errors. Ford related the story, which may or may not be apocryphal, that when AT&T contemplated the design of their key pad, they called several calculator companies, hoping they would share the research that led them to the opposite configuration. Much to their chagrin, AT&T discovered that the calculator companies had conducted no research at all. From our contacts with Sharp and Texas Instruments, two pioneers in the calculator field, it seems that this story could easily be true.

Terry L. Stibal, one of several readers who posed this Imponderable, suggested that if the lower numbers were on the bottom, the alphabet would then start on the bottom and be in reverse alphabetical order, a confusing setup. This might have entered AT&T's thinking, particularly in the "old days" when phone numbers contained only five digits, along with two exchange letters.

Submitted by Jill Gernand, of Oakland, California. Thanks also to: Lori Bending, of Des Plaines, Illinois, and Terry L. Stibal, of Belleville, Illinois.

What Is the Difference Between a "Kit" and a "Caboodle"?

Anyone who thinks that changes in the English language are orderly and logical should take a look at the expression "kit and caboodle." Both words, separately, have distinct meanings, but the two have been lumped together for so long that each has taken on much of the other's meaning.

Both words have Dutch origins: "Kit" originally meant tankard, or drinking cup, while "Boedel" meant property or house-

hold stuff. By the eighteenth century, "kit" had become a synonym for tool kit. For example, the knapsacks carried by soldiers that held their eating utensils and nonmilitary necessities were often called "kits." "Boodle" became slang for money, especially tainted money. By the nineteenth century, "caboodle" had taken on connotations of crowds, or large numbers.

Yet the slurring of meanings occurred even before the two terms became inseparable. The *Oxford English Dictionary* quotes from Shelley's 1785 *Oedipus Tyrannus*, "I'll sell you in a lump the whole kit of them." In this context, "caboodle" would seem more appropriate than "kit."

By the mid-nineteenth century, "kit" had found many companion words in expressions that meant essentially the same thing: "kit and biling"; "whole kit and tuck"; "whole kit and boodle" and "whole kit and caboodle" were all used to mean "a whole lot" or "everything and everyone." The *Dictionary of Americanisms* cites a 1948 *Ohio State Journal* that stated: "The whole caboodle will act upon the recommendation of the *Ohio Sun*."

The expression "kit and caboodle" was popularized in the United States during the Civil War. The slang term was equally popular among the Blue and the Gray. Although the expression isn't as popular as it used to be, it's comforting to know that old-fashioned slang made no more sense than the modern variety.

What Is the Purpose of the Ball on Top of a Flagpole?

We were asked this Imponderable on a television talk show in Los Angeles. Frankly, we were stumped. "Perhaps they were installed to make the jobs of flagpole sitters more difficult," we ventured. "Or to make flagpole sitting more enjoyable," countered host Tom Snyder. By turns frustrated by our ignorance and outwitted by Mr. Snyder, we resolved to find the solution.

According to Dr. Whitney Smith, executive director of the Flag Research Center in Winchester, Massachusetts, the ball may occasionally be combined with a mechanism involved with the halyards that raise and lower a flag, but this juxtaposition is only coincidental. Much to our surprise, we learned that the ball on top of a flagpole is purely decorative.

Actually, the earliest flaglike objects were emblems—an animal or other carved figure—placed atop a pole. Ribbons beneath these insignia served as decoration. According to Dr. Smith, the importance of the two was later reversed so that the design of the flag on a piece of cloth (replacing the ribbons) conveyed the message while the finial of the pole became ornamental, either in the form of a sphere or, as the most common alternatives, a spear or (especially in the United States) an eagle.

George F. Cahill, of the National Flag Foundation, believes that a pole just isn't as pleasing to the eye without something on top. Spears don't look good on stationary poles, and eagles, while visually appealing, are more expensive than balls or spears. Cahill adds another advantage of the ball: "On poles that are carried, a spear can be a hazard, not only to individuals, but to woodwork and plaster, and eagles are cumbersome and easily breakable. So, the ball gives the pole a safe and rather attractive topping and finish."

We speculated that perhaps birds were less likely to perch on a sphere than a flat surface, thus saving the flag from a less welcome form of decoration. But Cahill assures us that birds love to perch on flagpole balls.

We may never have thought of these balls as aesthetic objects, but *objets d'art* they are.

Why Does Wayne Gretzky Wear a Ripped Uniform?

Hockey players don't tend to make as much money as, say, top basketball players, but surely the biggest star in hockey can afford an unmangled uniform. Can't he?

Actually, Gretzky's uniform isn't ripped, and he is not trying to affect a "punk" look. Gretzky always tucks one corner of his shirt into the back of his pants, which only makes his shirt *appear* to be torn. The story of how Gretzky started this practice is a fascinating one, told to us by the National Hockey League's Belinda Lerner:

> It began when Wayne was a young hockey player competing with older boys. His uniform shirt, if left out of his pants, would hang down to his knees. Wayne tucked the shirt inside his pants to give a taller appearance. Now Wayne, who is six feet tall, tucks in the corner of his shirt out of habit and superstition. Bill Tuele, Public Relations Director for the Edmonton Oilers, tells me Wayne has a piece of Velcro sewn into his pants so the shirt is securely fastened throughout the game.

Submitted by Lorin Henner, of New York, New York.

Why Is There Always Pork in Cans of Pork and Beans? Does That Tiny Little Hunk of Fat, Which Is Presumably Pork, Really Add Flavor? It's Disgusting to Look at, So Why Do They Put It In? Why Not "Lamb and Beans" or "Crickets and Beans"? Why Always Pork?

Perhaps it will comfort you to know that yes, indeed, the pork is placed into the can for flavor. Pork and beans are actually cooked in the can. One fairly large piece of pork is placed in the can before cooking. After being heated during processing, it melts down to the size you see in the can, its flavor having permeated the beans.

We spoke to Kathy Novak, a Consumer Response Representative at Quaker Oats, the parent company of Stokely-Van Camp, who told us that they receive quite a few inquiries about the pork from fans of Van Camp's Pork and Beans, including more than a few angry missives from those who opened a can that inexplicably did not contain the piece of pork. So there is no doubt that the pork does have its fans. James H. Moran, of Campbell's Soup Company, says that many of his company's customers eat the pork, while others do not.

Do manufacturers have to include a piece of pork to call the product "Pork and Beans"? Not really. Some producers use rendered pork liquid instead of a solid piece of meat, and are legally entitled to call their product "Pork and Beans."

Submitted by Joel Kuni, of Kirkland, Washington.

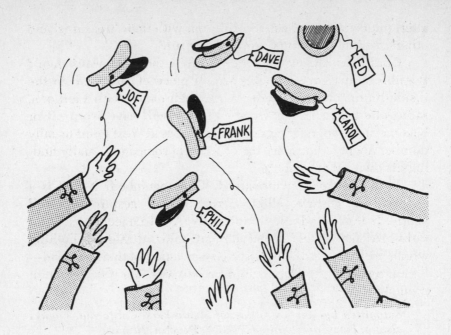

How Do Military Cadets Find Their Caps After Tossing Them in the Air upon Graduation?

Be it West Point, Annapolis, or Colorado Springs, the tradition is the same: at the end of graduation ceremonies, after the class is called to attention for the last time and the immortal words "You are dismissed" are uttered, the former cadets fling their caps in the air. Occasionally, hats will fly at sporting events as well. How are they retrieved?

The press and relatives grab a few. But the vast majority of the caps are claimed by children. Lieutenant Colonel James A. Burkholder, Commandant of Cadets at the U.S. Air Force Academy, wrote that after most, but not all, of the graduates throw their hats in the air, "children under 12 are allowed to scramble to get [the hats]. It becomes 'finders keepers.' Keeping the children off the field prior to that moment is also a sight to see. Thus, after graduation, you will see children with their 'treasures' and others, without hats, in all sorts of despair."

Could the cadets find their caps if they did want them? Possibly. Caps have a pocket with a piece of cardboard in the inside lining, on which cadets write their names with a felt pen. More often than not, however, the ink will have worn off or become smeared. As the graduating classes at West Point usually number about a thousand, the chances of someone actually finding his own hat are remote.

Is the hat tossing rehearsed? Choreographed? No. It is a spontaneous gesture, albeit a spontaneous gesture repeated yearly. Is it frowned upon? Not really. As Al Konecny, Assistant Public Affairs Officer at West Point, told us, there is nothing wrong with the graduates tossing away a part of their uniform— it's no longer their proper uniform, anyway. They've just been promoted!

Submitted by Merry Phillips, of Menlo Park, California. Thanks also to: Paul Funn Dunn, of WSOY, Decatur, Illinois.

Why Does American Electricity Run on A.C. Rather Than D.C.?

Direct current flows only in one direction. Alternating current flows back and forth continuously. Thomas Edison was a proponent of direct current (he had a financial stake in it), which worked fine in the early days of electric light, because the generators were very close to the lights that used electricity.

But as demand for electric light increased, D.C. proved inadequate. Electric current loses the least energy when traveling at high voltages. It was then uneconomical to transform D.C. to the high voltages necessary for long-distance transmissions. Direct current circuits would have required generating stations every three or four miles, unfeasible in the sprawling United States.

In 1885, a young man named George Westinghouse bought the U.S. patent for alternating current from inventor Nikola Tesla. Not only could A.C. transmit higher voltages more cheaply than D.C., the voltages could be raised or lowered by switching only one transformer. With its relative flexibility and lower cost, A.C. quickly became the U.S. standard.

Submitted by Larry Hudson, of Nashville, Tennessee. Thanks also to: John Brandon, of Davis, California.

What Is That Sniffing Noise Boxers Make When Throwing Punches?

Listen carefully to any boxing match, or to any boxer shadow-boxing, and you will hear a sniffing sound every time a punch is thrown. This sound is known to many in the boxing trade as the "snort."

A "snort" is nothing more than an exhalation of breath. Proper breathing technique is an integral part of most sports, and many boxers are taught to exhale (usually, through their nose) every time they throw a punch. Scoop Gallello, president of the International Veteran Boxers Association, told *Imponderables* that when a boxer snorts while delivering a punch, "he feels he is delivering it with more power." Gallello adds: "Whether this actually gives the deliverer of the punch added strength may be questionable." Robert W. Lee, president and commissioner of the International Boxing Federation, remarked that the snort gives a boxer "the ability to utilize all of his force and yet not expend every bit of energy when throwing the punch. I am not sure whether or not it works, but those who know much more about it than I do continue to use the method and I would tend to think it has some merit."

The more we researched this question, the more we were struck by the uncertainty of the experts about the efficacy of the snorting technique. Donald F. Hull, Jr., executive director of the International Amateur Boxing Association, the governing federation for worldwide amateur and Olympic boxing, noted that "While exhaling is important in the execution of powerful and aerobic movements, it is not as crucial in the execution of a boxing punch, but the principle is the same." Anyone who has ever watched a Jane Fonda aerobics videotape is aware of the stress on breathing properly during aerobic training. Disciplines as disparate as weightlifting and yoga stress consciousness of inhalation and exhalation. But why couldn't any of the boxing experts explain why, or if, snorting really helps a boxer?

Several of the authorities we spoke to recommended we contact Ira Becker, the doyen of New York's fabled Gleason's Gymnasium, who proved to have very strong opinions on the subject of snorting: "When the fighter snorts, he is merely exhaling. It is a foolish action since he throws off a minimum of carbon dioxide and some vital oxygen. It is far wiser to inhale and let the lungs do [their] own bidding by getting rid of the CO_2 and retaining oxygen."

The training of boxing, more than most sports, tends to be ruled by tradition rather than by scientific research. While most aspiring boxers continue to be taught to snort, there is obviously little agreement about whether snorting actually conserves or expends energy.

Why, in Any Box of Assorted Chocolates, Are the Caramels Square, the Nougats Rectangular, the Nuts Oval, and the Creams Circular?

Before we are inundated with letters, let's square away one fact. Not all chocolatiers conform to the geometric code stated in the Imponderable above. But most do. Who invented the scheme?

Despite contacting all of the biggest chocolate makers in the United States, and some of the older smaller ones, we couldn't come up with a definitive answer. Whether it was the venerable Thompson Candy Company in Meriden, Connecticut, See's Candies in South San Francisco, California, or Whitman's, all simply said that these shapes were "traditional."

When most chocolates were dipped by hand, many companies put an easily identifiable code on each chocolate, usually in the form of a swirly script on top of each piece. At Whitman's, for example, the code ran like this:

> A "V" usually signified a vanilla cream center; flat-topped chocolates with an open "C" often indicated a chocolate butter cream center; dome-shaped chocolates inscribed with a closed "C" would reveal a cherry cordial. Square-shaped chocolates with a "V" indicated a vanilla caramel center. An "O" stood for an orange cream; "P" for pineapple cream; and "R" for raspberry cream.

In the past, these fillings were coated with chocolate by a person who quickly dipped the centers in hot melted chocolate and then set aside the pieces to cool and solidify. Most boxed centers today are coated with an automatic enrober. Charlotte H. Connelly, of Whitman's, describes the process:

> In the enrobing process, candy centers are arranged on a moving belt which passes over a pool of chocolate which coats the underside. Then the centers move under a curtain of chocolate which coats each piece with precisely the right amount of melted chocolate. . . . In many instances, with the demise of the hand-dipper, the individual codes have ceased to exist. Although the shapes are

used by many confectionary manufacturers, this is certainly not industrywide as many boxed chocolates suppliers do not conform to these patterns.

How right she is. Most of Fanny Farmer's creams, for example, are circular, but their cream caramels and walnut creams are square. See's Candies produces approximately 130 different pieces of confection. And although each piece does have an individual marking, only the most diehard customer could possibly commit them all to memory.

What the world is clamoring for is a visual guidebook to chocolate centers, to stop the needless despair caused when an innocent person selects a nougat when he thinks he has chosen a caramel. Excuse us, while we call our editor. . . .

Submitted by Mrs. Marjorie S. Fener, of Hempstead, New York.

Whatever Happened to Pay Toilets?

Going to the bathroom is one of the few activities that has gotten cheaper of late. Pay toilets used to be the rule in airports and bus and train stations, and one would often encounter them in gas stations and restaurants.

Pay toilets were never meant to be profit-making enterprises, but merely a method to help defray the costs of cleaning the bathrooms. It was presumed that the dime or quarter "entrance fee" would motivate users to keep the pay stalls cleaner.

It didn't work, though, for instead of encouraging users to exercise best behavior, bathrooms with pay toilets were often trashed by angry patrons.

The vast majority of pay toilets in the 1950s and 1960s were operated by municipalities. According to Ben Castellano of the Federal Aviation Administration, the small amount of revenue generated by pay toilets in airports simply was not worth the attendant hassles: the numerous complaints about their presence and the constantly broken locks that rendered toilets unusable.

But the real death knell of the pay toilet came with several lawsuits filed against municipalities by women's groups. Pay toilets were sexually discriminatory, they argued, because women, unlike men, were forced to pay to urinate. Instead of putting women on the honor code or installing human or video monitors, most cities relented and abandoned the pay toilet. Even male chauvinists were forced to admit that the women's movement had struck a blow for humankind.

When a Pothole Is Formed on the Road, Why Don't We See the Displaced Concrete?

Of course, the endless procession of automobiles and trucks weakens concrete, but the real culprit in the creation of potholes is moisture. Potholes are usually formed when moisture penetrates the pavement. Combined with the destructive effects of alternating cycles of freeze and thaw, especially in the spring, and various chemicals, moisture weakens the concrete from within.

Eugene W. Robbins, president of the Texas Good Roads/Transportation Association, gave us the most succinct explanation of what happens to the "missing" concrete: "When the pothole forms, the material is broken loose, is pulverized or thrown to the side of the road by vehicle tires, and is blown away by wind or washed away by rain."

Submitted by Chuck Appeldoorn, of Woodbury, Minnesota.

Why Do Most Cities in the United States Put a Maximum-Height Restriction on a Fence a Homeowner May Put Around His Residence, and How Do They Decide on the Maximum?

Our *Imponderables* research team looked into this pesky problem and found that virtually all cities have local ordinances defining maximum heights for fences in residential neighborhoods. Are these restrictions fussy and capricious?

Before we can answer this Imponderable, we must not shrink from asking, without trepidation, the Big Question: what is the purpose of a fence? There are at least six common reasons why homeowners erect fences:

1. To block visual access to their property
2. To demarcate property lines
3. To inhibit access by unwanted people or animals
4. To protect property from the elements
5. To improve climatic conditions (particularly, to break up strong winds)
6. To enhance the appearance of the property

We can all agree, then, that the fence is a noble and worthwhile institution. So why can't homeowners build them to their preferred height specifications?

Because fences also cause problems. Enormous problems. And problems relating to their height are high on the list, all considerations in the decision to codify height restrictions:

1. *High fences are a safety hazard.* This is the number one reason why most localities enact restrictive ordinances. High fences at intersections or near driveways obstruct the vision of motorists and pedestrians. Fences of more than three feet near driveways are particularly dangerous, for they block the view of small children. Most cities also severely restrict the height of fences in the front of corner houses.

2. *As any regular viewer of* The People's Court *or regular reader of Ann Landers can testify, neighbors constantly fight about fences.* They fight about how high they are, what color they are, whose property they are on, and whose responsibility it is to maintain them. Many fences are built solely to irritate neighbors; these are called "spite fences." Without zoning ordinances, municipal governments were without the means to settle such disputes.

3. *Fences can block or obstruct the view, the available light, and the air flow of adjoining properties.* Just as the construction of a skyscraper can totally disrupt the surrounding en-

vironment, so can a relatively low fence in a residential neighborhood.

4. *Many people feel that fences are ugly.* But try to argue this point with a neighbor who has built a fence solely to secure more privacy.

5. *Fences can change the feel of a neighborhood.* Nothing warms up Elm Street like a nice six-foot high barbed wire topped electric chain-link fence.

6. *The same fence that diminishes wind in the winter blocks out cool breezes for the homeowner AND HIS NEIGHBOR during the summer.* The next-door neighbor becomes a passive victim of the fence.

You can imagine, then, the quandary that the city planner faces when trying to determine the proper height levels for residential fences. Zoning ordinances must regulate not only the height of fences, but their degree of openness, which materials can be used to construct fences, how a fence is defined, and how these regulations are to be enforced.

Laws must be enacted to allow homeowners to build fences without receiving permission from abutting neighbors, or the city risks needless delays and squabbles. And laws must be written to provide for exceptions. Fences around tennis courts, for example, must be built higher than other residential fences or there will be safety risks to neighbors and some rather angry tennis players. In mixed-use zones, the law must discriminate between commercial and residential property. The risk of children or criminals climbing into dangerous factories or near hazardous substances far outweighs the aesthetic damage done by a high fence.

So have a little compassion for these zoning ordinances, as arbitrary as they may seem. Sheaves of material sent by the American Planning Association indicate that planners are trying to give you a break. For example, in most localities, fences in side- and backyards are allowed to be higher than in front yards, because high fences in less public territory pose less of a safety risk.

Corner houses are likely to face the most restrictions. In Salt Lake City, for example, fences within thirty feet of an intersection or on any corner lot can be built no higher than three feet. A Coral Gables, Florida, ordinance restricts walls or fences to three feet high if they are within twenty feet of a street or alley intersection. Other cities also restrict any other structure, man-made or natural, from blocking visual access near an intersection.

As crime rates have escalated in surburban as well as urban neighborhoods, many cities have raised their height limits, often from four to six feet. Many residents feel more secure with a high fence around the perimeter of their property. Los Angeles has even considered upping its maximum height from six to eight feet for this very reason.

While the current trend is toward higher fences, don't wait for deregulation to hit local ordinances. High fences are still considered by most property associations to be eyesores and can adversely affect the property value of a single home or a whole neighborhood. When the subject of property values rears its ugly head, other considerations often have a way of fading from view.

Submitted by Bert Sailer, of Los Angeles, California.

Why Do Your Feet Swell Up So Much in Airplanes?

We talked to two specialists in aviation medicine who assured us that there is no reason why atmospheric changes in airplanes would cause feet to swell. Both assured us that the reason your feet swell up on a plane is the same reason they swell up on the ground—inactivity.

Your heart is not the only organ in the body that acts as a pump; so do the muscles of the legs. Walking or flexing a leg muscle assists the pumping effect. On a plane, you are not only confined in movement but sitting with the legs perpendicular to the floor. If you sit for prolonged periods without muscular activity, blood and other fluids collect in the foot with the assistance of gravity.

It doesn't really matter whether you leave your shoes on or off during periods of inactivity. If left on, they will provide external support, but they will inhibit circulation, feel tight—and will not prevent feet from swelling, in any case. If you take your shoes off, you will feel more comfortable, but you'll have a tough time putting your shoes back on, and most of us don't take our shoehorns along on planes.

The pooling of fluids in the feet can happen just as easily in a bus, a train, or an office. Most people's feet swell during the day, which is why the American Podiatric Association recommends buying shoes during the middle of the afternoon. Many people require a shoe a half size to a full size larger in the afternoon than when they wake up.

If your feet swelling becomes a problem, consider airplane aerobics. A few laps around a wide-body plane will do wonders for your feet and will build up your appetite for that wholesome and delicious airplane meal that awaits you.

Submitted by Christal Henner, of New York, New York.

Why Are Hamburger-Bun Bottoms So Thin?

An irate caller from Champaign, Illinois, hit us with this Imponderable, and we immediately empathized. Who hasn't taken a juicy burger off the barbecue, placed a Bermuda onion, some ketchup, maybe a dollop of mustard on the patty, only to find the bun bottom wilting in his fingertips? The bun top only gets mangled by the fingertips of the eater. The bottom carries the weight of the burger, the bun, the eater's fingers, plus the grease from the meat itself. Halfway through eating the sandwich, the bun top looks like Grace Kelly; the bun bottom looks like Sam Kinison.

All of the commercial retail bakeries we spoke to were eager

WHY DO CLOCKS RUN CLOCKWISE?

to fault the slicers. Of course, all commercial slicers come with adjustable blades, so the bun *could* be sliced at any height. Surely, not all hamburger buns are missliced, so why don't they slice them higher?

We did find some explanations for the mystery of the puny bun bottom. Most hamburger buns purchased in the grocery store are approximately one and a quarter inches high. They are baked in molds that are half an inch high. Many bakeries slice the bun at this half-inch mark rather than the true midpoint. This is necessary because the tops of the hamburger buns puff up (three-quarters of an inch above the mold), and the top is relatively fragile. The lowest point of this fragile area is called the "shred line." Slices made above the shred line tend to be less clean. One of the reasons that the bottoms of McDonald's Big Mac buns stand up so well is that they are baked in one-inch molds, producing a higher shred line, so that they can be sliced at a proportionately higher point.

The hamburger-bun tops are, in fact, much more important from a marketing point of view. Nothing pleases browsers at the bakery shelf more than buns with a pronounced mushroom top. Because the mushroom top is above the shred line, bakers must decide between good looks and practicality when it comes time to slice the buns—guess which one they choose.

B. W. Crosby, of Pepperidge Farm, adds that "The flavor and texture of the bun comes from the top; therefore, the top needs to be substantial in size." Although the ingredients on the top of the bun are identical to those on the bottom, there is some substance to this argument. The sugar in the dough caramelizes on the top of the bun during baking (that's why the top of the bun is browner than the rest), adding some flavor to what is essentially an exceptionally bland product. Any other flavor enhancements, such as sesame seeds or onions, are also generally loaded on the top of the bun.

One bun expert we spoke to, Bill Keogh of American Bakeries, offered a unique and practical solution to the bun bottom crisis. When he eats a large, juicy hamburger and anticipates a potential problem, he simply turns over the hamburger and eats it bun-bottom up. The so-called bun top, now on the bottom, easily soaks up any footloose grease, for not only is it heavier, it is also wider; thus he simultaneously solves the common problem of trying to eat a sandwich when the patty is wider than the bun bottom. This type of ingenuity is what separates us from the anthropoids.

Why Do Golfers Yell "Fore" When Warning of an Errant Golf Shot?

This expression, popularized by former President Gerald Ford, actually started as an English military term. When the troops were firing in lines, the command " 'ware before" indicated that it might be prudent for the front line to kneel so that the second line wouldn't blow their heads off.

"Fore" is simply a shortened version of the "before" in " 'ware before."

Submitted by Cassandra A. Sherrill, of Granite Hills, North Carolina.

Why Are All Executions in the United States Held Between Midnight and Seven A.M.?

Executions in the United States were not always held in the wee hours. Until the 1830s, most executions were hangings, public affairs that were usually performed at noontime in town squares. Government and penal officials generally believed that public executions acted as a deterrent to would-be felons, so they timed them to attract the largest crowd possible.

Public executions were not without problems even for proponents of capital punishment. Condemned criminals were allowed to give a last statement, which often turned into a tirade against the government and the church. The public execution gave often deranged individuals an opportunity to mock the values and institutions that the justice system was designed to protect. Some criminals successfully played upon the sympathy of the crowd, portraying themselves as martyrs.

About this time, opposition to public executions began to be heard, with both philosophical and practical objections being raised. Many found public executions barbaric, especially because the crowd more resembled crazed football fans than witnesses to a human death. Local merchants in big cities usually disliked public executions because they disrupted business (for the same reason, small-town merchants liked public executions —they attracted potential customers).

Sociologist Richard Moran of Mount Holyoke College, a leading expert on the history of capital punishment, stresses that throughout American history, those against it have sought public executions, believing that if the American public were exposed to the barbarity of executions, it would reject them.

Those in favor have always sought to make executions private. This trend was first manifested in the 1830s, when several states decided to perform executions within prison walls. Rhode Island was the first state to abolish public executions, in 1833, and Pennsylvania followed the year after. Actually, the public could still attend hangings inside prison, but only by purchasing tickets. During this period, most executions were still held during the afternoon.

Slowly, many states began to move their execution times to late evening, midnight, and even predawn hours. The practice of execution at dawn dates back centuries, when the military of many countries had firing squads execute the condemned as soon as there was enough light for the gunmen to see their target. Professor Moran, however, provides another reason for the early-morning hours: concealment. Even in the nineteenth century, most prison and government officials tried to diminish press coverage of executions, in order to minimize protest. Executions were held in the early morning so that the press couldn't print stories about the execution in that morning's newspaper. If they chose to cover the execution at all, the story had to appear on the following day, when it was likely to receive less prominence.

Some of the penal authorities we spoke to stressed the practical advantages of early-morning executions. Anthony P. Travisono, executive director of the American Correctional Association, told *Imponderables* that between midnight and six A.M., "there is very little activity at the institution and all is fairly quiet." He added, rather eerily, "also, the surge of power is stronger."

Professor Moran also suggests that prisoners are more disorderly on days of executions, so that performing executions while the inmates sleep is prudent.

We asked Moran if early-morning executions could be an expression of collective, if unconscious, shame about the whole enterprise. Although he didn't take a definitive stand on the issue, Moran acknowledged the possibility and bolstered the contention with an interesting fact about electrocutions. Not only are today's executions indoors and closed to the public, but electrocutions, without exception, are performed in rooms without windows. The grim task of taking a human life is accomplished without any natural light entering the chamber. We used to make executions a public ritual, symbolic of our repudiation of the criminal act and our affirmation of the need for justice. We now perform executions privately, almost furtively—as if we were the criminals.

Why Do Ants Tend to Congregate on Sidewalks?

With the help of several entomologists and pest control experts, we pieced together several reasons for this phenomenon.

1. Some species, particularly one actually called "pavement ants," prefer to nest on sidewalks and under rocks and other hard surfaces.

2. As John J. Suarez, technical manager of the National Pest Control Association, so elegantly put it: "Sidewalks are a favorite place for people to drop candy, fast food, food wrappers and soft drink containers." Ants are known for their industriousness, but they aren't dumb. If they are given offerings that require no effort on their part, they won't decline the largess.

3. Ants release pheromones, a perfume trail left from the nest to food sources. Pheromones are easily detected on sidewalks, which, as we have already learned, are often repositories

for food. Ants on sidewalks, then, are often merely picking up the scent left by scouts before them.

4. Sidewalks absorb and store heat. Ants run around naked. They prefer warmth.

5. The most popular explanation: darkish ants are more easily visible in contrast to the white sidewalk than on grass or dirt. Suarez speculates that the greater warmth of sidewalks may make the ants more active as well as more visible. But don't assume that because you can't see ants on your front lawn they aren't there. THEY ARE EVERYWHERE.

Submitted by Daniel A. Placko, Jr., of Chicago, Illinois.

Why Do American Cars Now Have Side-View Mirrors on the Passenger Side with the Message, "Objects in the Mirror Are Closer Thán They Appear"?

A reasonable person might ask why the American automobile industry had to "improve" on those hopelessly old-fashioned side-view mirrors that didn't distort one's perception of distance. And why, if the inside rear-view mirror shows objects without distortion, can't the side mirror do the same?

Car manufacturers are required to provide flat, unit magnification mirrors on the driver's side of the car. The driver-side mirrors offer the same undistorted image as the mirror in your medicine cabinet.

The new mirrors are convex (for those who forgot their high-school science, convex surfaces curve outward, as opposed to a spoon, which has a concave surface). Convex mirrors have one huge advantage over flat mirrors—they allow a much wider angle of vision. Engineers have found that convex side-view mirrors afford drivers a much clearer view of the passenger side of the car than the old combination of rear-view mirror and conventional side-view mirror. The rear-view mirror, if used alone, leaves blind spots that can lull drivers into complacency when they are considering making lane changes. Drivers are less likely to be sideswiped when consulting a wide-angle side-view mirror, even if an oncoming car is closer than it appears, because they are more likely to spot the car in the first place.

The immortal words, "Objects in the mirror are closer than they appear," are mandated by federal law on all convex mirrors. The government has also set specific standards for the curvature of convex mirrors. The average radius of the curvature for convex mirrors should be no less than thirty-five inches and no greater than sixty-five inches.

Ed Stuart, a representative of Chrysler Motors, told *Imponderables* that the convex mirrors are particularly popular with freeway and turnpike drivers, who can see oncoming cars streaming in from entry ramps much more easily. The biggest danger of the convex mirror is that because objects in the mirror are closer than they appear, drivers will think they have more room to pass another car than they really do. But most drivers look through the undistorted rear-view mirror rather than the side-view mirror before making a lane change anyway, and the prudent driver should check over his shoulder before making his move.

Submitted by Loretta McDonough, of Richmond Heights, Missouri.

Why Do Dogs Smell Funny When They Get Wet?

Having once owned an old beaver coat that smelled like a men's locker room when it got wet, we assumed that the answer to this Imponderable would have to do with fur. But all of the experts we spoke to agreed: the funny smell is more likely the result of dogs' skin problems.

First of all, not all dogs do smell funny when they get wet. Shirlee Kalstone, who has written many books on the care and grooming of dogs, says that certain breeds are, let us say, outstanding for their contribution to body odor among canines. Cocker spaniels and terriers (especially Scotties) lead the field, largely because of their propensity for skin conditions. (Cockers, for example, are prone to seborrhea.) Jeffrey Reynolds, of the National Dog Groomers Association, adds that simple rashes and skin irritations are a common cause of canine body odor, and that water exacerbates the smell. In his experience, schnauzers are particularly susceptible to dermatological irritations.

Of course, dogs occasionally smell when they get wet because they have been rolling in something that smells foul. Gamy smells are usually caused by lawn fertilizer, for example.

Regular grooming and baths can usually solve the odor problem, according to Kalstone. Don't blame the water, in other words—blame the owner.

Submitted by Robert J. Abrams, of Boston, Massachusetts.

Why Do All Dentist Offices Smell the Same?

You are smelling what Dr. Kenneth H. Burrell, assistant secretary of the Council on Dental Therapeutics, calls the "mixture of essential oils dentists frequently use in the course of treatment." Many of these oils are natural or synthetic derivatives of products found in your household. A number of dental medicaments contain camphor, the same pungent substance that repels moths, while others are derivatives of items found on the spice rack: thyme and clove.

The most prominent scent in the dentist's office, the one that makes you claim that all dentist offices smell alike, is probably eugenol, a colorless or pale liquid that is the essential chemical constituent of clove oil. Dentists combine solutions of clove oil or eugenol with a mixture of rosin and zinc oxide to prepare a protective pack after gum surgery or as a temporary cement. Eugenol is also used as an antiseptic, especially in root-canal therapy, and as an anodyne (painkiller). This versatile liquid is also part of the mixture for temporary fillings, impression materials, and surgical dressings after periodontal work.

We at *Imponderables* have pondered of late why clove candies and gum have come and gone. Our pet theory is that rejection of this once-popular flavor comes from a generation of sense memories unconsciously associating cloves with the dentist's office.

Submitted by Julie Lasher and Brian Scott Rossman, of Sherman Oaks, California.

What Are Those Large Knobs Between Sets of Escalators in Department Stores?

The knobs' sole purpose in life is to keep miscreants from sliding down the flat space between the escalators—or at least to guarantee that if misceants *do* slide down in between the escalators, they'll have a bumpy ride.

At Westinghouse Electric Corporation, the knobs are known as an "Anti-Slide Device" and are used on the wide decking between escalators or between an escalator and a wall. As Westinghouse engineer Robert L. Meckley points out, the knobs not only prevent rowdy kids from sliding down, but also keep purses and other baggage (such as Baccarat crystal) from flying down and crashing on the floor. Although the Anti-Slide Devices are hardly high-tech, they do the job.

Submitted by Liz Sblendorio, of Hoboken, New Jersey.

Why Is Jack the Nickname for John?

Believe it or not, a whole book has been written on this subject: *The Pedigree of Jack and of Various Allied Names* by E. W. B. Nicholson. (Don't look for it in your bookstore; it was published in 1892.)

The history of Jack as a pet name for John is a long and tangled one, as these things usually are. Most people assume that Jack is derived from the French Jacques, and that Jack should therefore be short for James rather than John. Nicholson debunked this notion, claiming that there is no recorded example of Jack ever being used to represent Jacques or James.

Jack is actually derived from the name Johannes, which was shortened to Jehan and eventually to Jan. The French were fond of tacking the suffix *-kin* onto many short names. French nasalization resulted in the new combination being pronounced Jackin instead of Jankin. The name Jackin was shortened to Jack. The Scottish version, Jock, was a similar contraction of Jon and *-kin*.

By the fourteenth century, Jack had become a synonym for *man* or *boy*, and later was also used as a slang name for sailors (thus the Jack in Cracker Jack).

In the mid-nineteenth century, Jack became popular as a Christian name, and it remained so until its use peaked in the 1920s. At that point, the diminutive Jackie became popular, propelled by child stars Jackie Cooper and Jackie Coogan. The feminine equivalent, Jacqueline, became the rage in the 1930s, and Jackie, for a short period, became a unisex name. Jack never regained its prominence, though there was a small surge after the United States elected a popular president named John, whose pet name was Jack.

Submitted by Michael Jeffreys and Krissie Kraft, of Marina del Rey, California.

Which Side Gets the Game Ball When a Football Game Ends in a Tie?

Jim Heffernan, director of public relations for the National Football League, told *Imponderables* that NFL rules require that each home team provide twenty-four footballs for the playing of each game. The home team and the visiting team each provide additional balls for their pregame practice.

A "game ball," contrary to popular belief, is not one football given to the winning side. Game balls are rewards for players and coaches who, as Heffernan puts it, "have done something special in a particular game." The game-ball awards are usually doled out by the coach; on some teams, the captains determine the recipients.

The same holds true in college football. James A. Marchiony, director of media services for the National Collegiate Athletic Association, says, "Game balls are distributed at the sole discretion of each team's head coach; a winning, losing or tying coach may give out as many as he or she wishes."

Submitted by Larry Prussin, of Yosemite, California.

Why Do Ketchup Bottles Have Necks So Narrow That a Spoon Won't Fit Inside?

Heinz has had a stranglehold on the ketchup business in the Western world for more than a century, so the story of ketchup bottle necks is pretty much the story of Heinz Ketchup bottle necks. Ironically, although Heinz ads now boast about the *difficulty* of pouring their rather thick ketchup, it wasn't always so.

When Heinz Ketchup was first introduced in 1876, it was considerably thinner in consistency. It came in an octagonal bottle with a narrow neck intended to help impede the flow of the product. Prior to the Heinz bottle, most condiments were sold in crocks and sharply ridged bottles that were uncomfortable to hold.

Over the last 111 years, the basic design of the Heinz Ketchup bottle has changed little. The 1914 bottle looks much like today's, and the fourteen-ounce bottle introduced in 1944 is identical to the one we now use. Heinz *was* aware that as their ketchup recipe yielded a thicker product, it poured less easily through their thin-necked bottle. But they also knew that consumers preferred the thick consistency and rejected attempts to dramatically alter the by-now-familiar container.

Heinz's solution to the problem was the marketing of a twelve-ounce wide-mouth bottle, introduced in the 1960s. Gary D. Smith, in the communications department of Heinz USA, told *Imponderables* that the wide-mouth bottle, more than capable of welcoming a spoon, is the "least popular member of the Heinz Ketchup family." He added, though, that "its discontinuance would raise much fervor from its small band of loyal consumers who enjoy being able to spoon on" their ketchup.

In 1983, Heinz unveiled plastic squeeze bottles, which not only solved the pourability problem but also solved the breakability problem. The sixty-four−ounce plastic size, while mammoth, still has a relatively thin neck.

Until 1888, Heinz bottles were sealed with a cork. The neckband at the top of the bottle was initially designed to keep a foil cap snug against its cork and sealing wax. Although it was rendered obsolete by the introduction of screw-on caps, the neckband was retained as a signature of Heinz Ketchup.

Submitted by Robert Myers, of Petaluma, California.

$\frac{99}{100}$, $\frac{98}{100}$, $\frac{97}{100}$, $\frac{96}{100}$ ...

Soap Vat
QUALITY CONTROL STATION

$\frac{44}{100}$
... or bust!

Ivory Soap Advertises Its Product as 99 and $^{44}/_{100}$ Percent Pure—99 and $^{44}/_{100}$ Percent *What?* And What Is the Impure $^{56}/_{100}$ Percent of Ivory Soap?

Procter & Gamble, in the late nineteenth century, sold many products made of fats, such as candles and lard oil, as well as soap. Ivory Soap was originally marketed as a laundry soap, but the company was smart enough to realize its product's potential as a cosmetic soap. The only problem was that most consumers were buying castile soaps (hard soaps made out of olive oil and sodium hydroxide) and considered laundry soap inappropriate for their personal grooming.

In order to convince consumers that its soap was wholesome, Procter & Gamble employed an independent scientific consultant in New York City to determine exactly what a pure soap was. The answer: a pure soap should consist of nothing but fatty acids and alkali; anything else was foreign and superfluous.

Samples of Ivory Soap were sent to the same chemist for analysis. Much to the manufacturer's surprise, Ivory, by the consultant's definition, was "purer" than the competing castile soaps—containing only 0.56 percent "impurities." The impurities, then and now, were rather innocent:

> Uncombined alkali 0.11 percent
> Carbonates 0.28 percent
> Mineral matter 0.17 percent

The first Ivory advertisement was placed in a religious weekly, *The Independent*, on December 21, 1881. Procter & Gamble decided to emphasize the positive, and right away hammered at their product's advantages. Ivory Soap was trumpeted as "99 and $^{44}/_{100}$ percent pure," a rare advertising slogan in that it has lasted longer than a century.

Submitted by Linda A. Wheeler, of Burlington, Vermont.

Why Do We Grow Lawns Around Our Houses?

At first blush, this Imponderable seems easily solved. Lawns are omnipresent in residential neighborhoods and even around multiunit dwellings in all but the most crowded urban areas. Lawns are pretty. Enough said.

But think about it again. One could look at lawns as a monumental waste of ecological resources. Today, there are approximately 55 million home lawns in the United States, covering 25 to 30 million acres. In New Jersey, the most densely populated state, *nearly one-fifth of the entire land area is covered with turfgrass,* twice as much land as is used for crop production. Although turfgrass is also used for golf courses and public parks, most is planted for lawns. The average home lawn, if used for growing fruits and vegetables, would yield two thousand dollars worth of crops. But instead of this land becoming a revenue generator, it is a "drainer": Americans spend an average of several hundred dollars a year to keep their lawns short and healthy.

If the purpose of lawns is solely ornamental, why has the tradition persisted for eons, when most conceptions of beauty change as often as the hem length of women's dresses? The Chinese grew lawns five thousand years ago, and circumstantial evidence indicates that the Mayans and Aztecs were lawn fanciers as well. In the Middle Ages, monarchs let their cattle run loose around their castles, not only to feed the animals, but to cut the grass so that advancing enemy forces could be spotted at a distance. Soon, aristocrats throughout Europe adopted the lawn as a symbol of prestige ("if it's good enough for the king, it's good enough for me!"). The games associated with lawns—bowls, croquet, tennis—all started as upper-class diversions.

The lawn quickly became a status symbol in colonial America, just as it was in Europe. Some homeowners used scythes to tend their lawns, but most let animals, particularly sheep, cows, and horses, do the work. In 1841, the lawn mower was introduced, much to the delight of homeowners, and much to the dismay of grazing animals and teenagers everywhere.

Dr. John Falk, who is associated with the educational research division of the Smithsonian Institution, has spent more time pondering this Imponderable than any person alive, and his speculations are provocative and convincing. Falk believes that our desire for a savannalike terrain, rather than being an aesthetic predilection, is actually a genetically encoded preference. Anthropologists agree that humankind has spent most of its history roaming the grasslands of East Africa. In order to survive against predators, humans needed trees for protection and water for drinking, but also grassland for foraging. If primitive man wandered away into rain forests, for example, he must have longed to return to the safety of his savanna home. As Falk commented in an interview in *Omni* magazine: "For more than ninety percent of human history the savanna was home. Home equals safety, and that information has to be fairly hard-wired if the animal is going to respond to danger instantaneously."

When we talked to Dr. Falk, he added more ammunition to support his theories. He has conducted a number of cross-cultural studies to ascertain the terrain preferences of people all over the world. He and psychologist John Balling showed subjects photographs of five different terrains—deciduous forest, coniferous forest, tropical rain forest, desert, and savanna—and asked them where they would prefer to live. The savanna terrain was chosen overwhelmingly. Falk's most recent studies were conducted in India and Nigeria, in areas where most subjects had never even seen a savanna. Yet they consistently picked the savanna as their first choice, with their native terrain usually the second preference.

Falk and Balling also found that children under twelve were even more emphatic in their selection of savannas, another strong, if inconclusive, indication that preference for savanna terrain is genetic.

In the *Omni* article, Falk also suggested that even the way we ornament our lawns mimics our East African roots. The ponds and fountains that decorate our grasses replicate the natural water formations of our homeland, and the popularity of umbrella-shaped shade trees might represent an attempt to re-create the acacia trees found in the African savanna.

Of course, psychologists have speculated about other reasons why we "need" lawns. The most common theory is that lawns and gardens are a way of taming and domesticating nature in an era in which affluent Westerners are virtually divorced from it. Another explanation is that lawns are a way of mapping territory, just as every other animal marks territory to let others know what property it is ready to defend. This helps explain why so many homeowners are touchy about the neighborhood kid barely scraping their lawn while trying to catch a football. As Dr. Falk told *Imponderables*, "People create extensions of themselves. When people create a lawn as an extension of themselves, they see a violation of their lawn as a violation of their space."

Lawns are also a status symbol, for they are a form of property that has a purely aesthetic rather than economic purpose. Historically, only the affluent have been able to maintain lawns —the poor simply didn't have the land to spare. Fads and fashions in lawns change, but there are usually ways for the rich to differentiate their lawns from the hoi polloi's. Highly manicured lawns have usually been the preference of the rich, but not always. In the Middle Ages, weeds were considered beautiful. In many parts of the world, mixed breeds of turf are preferred.

American taste has become increasingly conservative. Ever since World War II, the "ideal" American lawn has been a short, monoculture, weed-free lawn, preferably of Kentucky bluegrass. Falk sees these preferences as carry-overs from the technology used by American agronomists to develop grass for golf courses. Americans always want to build a better mousetrap; our "ideal lawn" has become just about the only type.

Americans have largely resisted the inroads of artificial grass. Although many team owners endorse it, sports fans by and large recoil at artificial turf in sports stadiums—perhaps another genetically determined predisposition.

Submitted by Rick Barber, of Denver, Colorado.

Why Do Many Exterminators Wear Hard Hats?

Our correspondent wondered why one of the largest exterminator companies, in its television commercials, dresses its exterminators with nice pants, a dressy shirt, and a hard hat. Is there any practical reason for the hard hat in real life? Is there a marketing reason?

The practical reason: pest-control operators often have to inspect crawl spaces, basements, and cellars full of obstacles— nails, heat ducts, spider webs, and other protruding objects from above. The hard hat helps reduce accidents.

The marketing reason: the hard hat conveys a professional image. Subliminally, the hard hat is supposed to make the customer think: "If the exterminator has to wear a hard hat, this work must be too dangerous for a civilian like me! Better leave it to the experts."

Submitted by Phil Feldman, of Los Angeles, California.

Who Was the Emmy That the Emmy Award Is Named After?

Not who, but what? Unlike the premier theater (Tony) and movie (Oscar) awards, the Emmy isn't named after a person.

In 1948, the president of the budding National Academy of Television Arts and Sciences, Charles Brown, formed a committee to select the outstanding achievements in television that year. He also asked for suggestions for a name and symbol for the award.

From the start, technological terms were the top contenders. "Iconoscope" (a large orthicon tube) was an early favorite, but the committee was afraid the name would be shortened to "Ike." "Tilly" (for television) was suggested, but cooler heads prevailed. Harry Lubcke, a pioneer television engineer and future president of the academy (1949–1950) offered "Emmy," a nickname for the image orthicon tube (state-of-the-art circuitry at that time), and it prevailed.

The statue itself was designed by Louis McManus, who received a gold lifetime membership in the academy and one of the six statuettes presented at the first Emmy Awards banquet on January 25, 1949. As McManus went up to receive his award, he is reputed to have been told, "Louis, here she is . . . our baby. She'll be here long after we're gone." Indeed, long after the image orthicon tube was gone.

Why Don't Dogs Develop Laryngitis, Sore Throats, Voice Changes, or Great Discomfort After Barking Continuously?

A caller on a talk show hit us with this Imponderable. The dog next door, left alone by his master, had been barking, continuously, for hours. Why didn't it hurt the dog's throat at least as much as the caller's ears?

We approached several vets and stumped some, but the consensus answer was best expressed by William E. Monroe, D.V.M., Diplomate, of the American College of Veterinary Internal Medicine:

> Dogs do occasionally get laryngitis and voice changes from excessive barking. It is not as common in dogs as in people because the motor control of the canine larynx (voice box) is not as refined as

that of humans for sound production. Therefore, the voice range is narrower and subsequent stress from phonation is probably not as severe. Since barking is not much a part of daily living for most pet dogs as speaking is for people, laryngitis manifested as a voice change is also not as frequently observed in dogs, even though it may be present.

Why Are There Eighteen Holes on a Golf Course?

In Scotland, the home of golf, courses were originally designed with varying numbers of holes, depending on the parcel of land available. Some golf courses, according to U.S. Golf Association Librarian Janet Seagle, had as few as five holes.

The most prestigious golf club, the Royal and Ancient Golf Club of St. Andrews, originally had twenty-two holes. On October 4, 1764, its original course, which had contained eleven holes out and eleven holes in, was reduced to eighteen holes total in order to lengthen them and make it more challenging. As a desire to codify the game grew, eighteen holes was adopted as the standard after the St. Andrews model.

What Does 0° in the Fahrenheit Scale Signify?

During our school days, we were forced to memorize various points in the Fahrenheit scale. We all know that the freezing point is 32° and that the boiling point is 212°. The normal human body temperature is the inelegantly unround number of 98.6°.

Countries that have adopted the metric system have invariably chosen the Celsius system to measure heat. In the Celsius scale, 0° equals the freezing point.

The Fahrenheit temperature scale was created by a German physicist named Gabriel Daniel Fahrenheit, who invented both the alcohol thermometer and the mercury thermometer. The divisions of his scale aren't quite as arbitrary as they might seem. Zero degrees was chosen to represent the temperature of an equal ice-salt mixture, and 100° was originally supposed to signify the normal body temperature. But Fahrenheit screwed up. Eventually, scientists found that the scale didn't quite work, and the normal body temperature was "down-scaled" to 98.6°.

Submitted by James S. Boczarski, of Amherst, New York.

What Does Each One-Degree Increment in the Fahrenheit Scale Signify?

Although his scale was not based on the freezing and boiling points, Fahrenheit recognized their significance. The interval between the boiling point (212°) and freezing point (32°) numbers exactly 180 degrees on the Fahrenheit scale, a figure with which scientists and mathematicians were used to working.

The increments in a temperature scale have no cosmic significance in themselves. The Celsius system, for example, is less precise than the Fahrenheit in distinguishing slight variations in moderate temperatures. Thus while 180 increments on the Fahrenheit scale are necessary to get from the freezing to the boiling point, the freezing point (0°) on the Celsius scale and the boiling point (100° C) are closer, only 100 increments apart.

In most cases, the meaning of the one-degree increments in temperature scales has more to do with what is intended to be measured by the scale than with any particular mathematical requirements. The Fahrenheit scale, intended for use in human thermometers, was designed originally to have 100°F represent the normal body temperature. Temperature scales now used by scientists, such as the Kelvin and Rankine scales, use absolute zero (the equivalent of −273.15° C or −459.67° F) as the base point. Rankine uses the same degree increments as Fahrenheit; Kelvin uses the Celsius degree.

Submitted by James L. Foley, of Calabasas, California.

Why Doesn't Rain Come Down the Chimney into the Fireplace When Smoke Can Get out of the Chimney?

Some residential buildings contain chimney caps, sloping structures that stand atop the chimney, as pictured below:

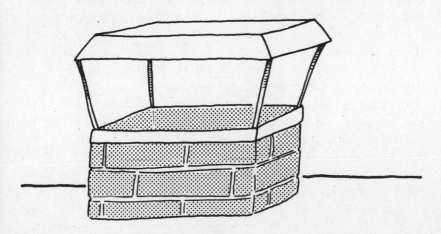

While the rain slides down the slopes, the smoke can easily escape under the cap.

But most buildings don't contain chimney caps and don't need them, for even large amounts of water can be absorbed by the bricks and masonry in a chimney. Indeed, the high absorbency of bricks is one of the reasons they are used in chimneys. In buildings of four or more stories, almost any quantity of water will be absorbed without the need of a chimney cap.

Submitted by Michael Jeffreys and Krissie Kraft, of Marina del Rey, California. Thanks also to: Leonard Scarpace, of Newhall, California.

Why Do Curad Bandage Wrappers Sparkle in the Dark When You Open Them?

Whenever we did promotion for the first volume of *Imponderables* and talked about why wintergreen Life Savers glow in the dark when you bite into them, we invariably received a phone call from someone asking why opening Curad wrappers caused the same phenomenon.

So we wrote to Colgate-Palmolive to unravel this mystery. The adhesive used to seal Curad bandage wrappers contains an ultraviolet dye. Most likely, the excitation and friction caused in the act of opening the wrapper causes the glow, which is visible only in the dark.

The research-and-development department of Colgate-Palmolive adds that static electricity might add to the sparkling effect. The sparkling is perfectly harmless and, if you are cut, a nice diversion from your pain.

Why Do Garment Labels Often Say "*Professionally* Dry-Clean Only"?

When was the last time you were propositioned by an *amateur* dry cleaner? Most folks don't take up dry-cleaning as an avocation, so when we were posed this Imponderable by a caller on the *Owen Spann Show*, we resolved to ferret out exactly who these dangerous amateur dry cleaners were.

It turns out that the veiled reference in "Professionally Dry-Clean Only" labels is not to incompetent practitioners, but to incompetent machines. What these labels are actually warning you against is cleaning the garments in the coin-op bulk dry-cleaning machines that used to be found frequently in laundromats. As much as anything, the labels are a convenient way for clothing manufacturers to avoid liability for the havoc the machines have wreaked. Molly Chillinsky, of the Coin Laundry Association, says that these bulk dry-cleaning devices are almost extinct; in time, the labels might become so as well.

Karen Graber, communications director of the International Fabricare Institute, the Association of [ahem] Professional Dry Cleaners and Launderers, adds that the Federal Trade Commission stipulates that along with the "Professionally Dry-Clean Only" warning, garment manufacturers should inform the consumer of any necessary modification in the basic dry-cleaning process. "Even the most professional dry cleaner might not know without being told that something should be dried at a low temperature, for example, or should not be pressed with steam." The clothing industry, unfortunately, often forgets to add these warnings.

Overcome by the spirit of Imponderables, Karen Graber added one of her own. Another clothing label warning that is sprouting up is the incantation: "Dry-clean only. Do not use petroleum or synthetic solvent." As there are only two kinds of solvent (you guessed it—petroleum and synthetic), her Imponderable is: what do you do with such a garment?

Graber's answer: "Leave it in the store, along with anything else you know from the label is bound to cause you and your dry cleaner some sleepless nights."

What Is the Difference Between "Flotsam" and "Jetsam"?

Although they sound suspiciously like two of Santa's missing reindeer, flotsam and jetsam are actually two different types of debris associated with ships. We rarely hear either term mentioned without the other close behind (and saying "jetsam" before "flotsam" is like saying "Cher" before "Sonny"). When we talk about "flotsam and jetsam" today, we are usually referring metaphorically to the unfortunate (for example, "While visiting the homeless shelter, the governor glimpsed what it is like to be the flotsam and jetsam of our society").

At one time, however, "flotsam" and "jetsam" not only had different meanings, but carried important legal disinctions. In English common law, "flotsam" (derived from the Latin *flottare*, "to float") referred specifically to the cargo or parts of a wrecked ship that float on the sea.

"Jetsam" (also derived from Latin—*jactare*, "to throw") referred to goods purposely thrown overboard in order either to lighten the ship or to keep the goods from perishing if the ship did go under.

Although the main distinction between the two terms was the way the goods got into the water, technically, to become jetsam, the cargo had to be dragged ashore and above the high-water line. If not, the material was considered flotsam, which included all cargo found on the shore between the high- and low-water lines.

Actually, two more terms, "lagan" and "derelict," were also used to differentiate cargo. "Lagan" referred to any abandoned wreckage lying at the bottom of the sea; "derelict" was the abandoned ship itself.

While insurance companies today have to pay out for flotsam, jetsam, lagan, *and* derelict, the old distinctions once dictated who got the remains. Jetsam went to the owner of the boat, but flotsam went to the Crown. The personal effects of nonsurviving crewmen could become flotsam or jetsam—depending on how far the debris traveled and whether it floated.

Why Do Doughnuts Have Holes?

The exact origins of doughnuts and their holes are shrouded in mystery and are a topic of such controversy that we have twice been caught in the middle of heated arguments among professional bakers on radio talk shows. So let us make one thing perfectly clear: we offer no conclusive proofs here, only consensus opinion.

Some form of fried cake has existed in almost every culture. "Prehistoric doughnuts"—petrified fried cakes with holes—have been found among the artifacts of a primitive Indian tribe. The Dutch settlers in America, though, are usually credited with popularizing fried cakes (without holes) in the United States, which they called "oily cakes" or *olykoeks*. Washington Irving, writing about colonial New York, described "a dish of balls of sweetened dough fried in hog's fat, and called dough nuts or oly koeks." Fried cakes became so popular in New York and New England that shops sprouted up that specialized in serving them with fresh-brewed coffee. In 1673, the first store-bought fried cakes were made available by Anna Joralemon in New York. Mrs. Joralemon weighed 225 pounds and was known affectionately as "the Big Doughnut."

The gentleman usually credited with the "invention" of the doughnut hole was an unlikely candidate for the job—a sea captain named Hanson Gregory. Supposedly, Captain Gregory was at the helm of his ship, eating a fried cake one night, when stormy weather arose. Gregory, needing both hands to steer the ship, spontaneously rammed the cake over one of the spokes. Impressed with his creation, Gregory ordered the ship's cook to make fried cakes with holes from then on.

Many other legends surround the creation of the doughnut hole. Plymouth, Massachusetts, advances the notion that the first doughnut hole was created when, in the seventeenth century, a drunken Indian brave shot an arrow through a kitchen window, punching out a piece of dough from the center of a cake just about to be fried. Pretty lame, Plymouth.

Regardless of the origin of the holes in doughnuts, we have learned that bakers disagree about its role in the making of a quality doughnut. Certainly, good doughnuts can be made without holes.

Thomas A. Lehmann, director of bakery assistance at the American Institute of Baking, told us that yeast-raised doughnuts can be made quite easily without the hole and points to the bismarck, or jelly-filled doughnut, as a perfect example. Lehmann adds, though, that if bismarcks were fried on the surface, the same way as conventional yeast-raised doughnuts, the holeless dough would tend to overexpand, turning into a ball shape. That is why most bakers prefer submersion frying, which results in a more uniform and symmetrical finished product.

"Cake" doughnuts, which are chemically leavened, can also be made without holes, but many experts believe that they lose their desired consistency without them. Glenn Bacheller, director of product marketing for Dunkin' Donuts, explains why the hole is important: "Heat does not penetrate the donut as readily [without the hole] and the interior of the donut tends to have a doughy texture. The only way to prevent this is to fry the donut longer, which results in the exterior of the donut being over fried."

Why Does a Newspaper Tear Smoothly Vertically and Raggedly Horizontally?

Newsprint is made up of many wood fibers. The fibers are placed on printers in pulp form, consisting of 80 to 90 percent water—the newsprint dries while in the machine. The printing machines are designed to line up the fibers in a horizontal position to add tear strength to the sheet vertically.

The basic purpose of lining up the fibers in one direction is simply to add stability to the sheet when the press is running. According to Ralph E. Eary, national director of production and engineering for the newspaper division of Scripps Howard, "All standard size newspapers are printed vertically on an unwound sheet of newsprint." A rip in one sheet endangers the whole printing process, and at best costs money and time.

In other words, the finished newspaper has a grain, just as a piece of meat or linen has a grain. (Even notice how hard it is to tear a bedsheet in one direction and how easy in another?) When you rip the newspaper vertically, you are tearing with the grain, or more accurately, between grains. The same principle is in effect when one consumes Twizzlers brand licorice. Individual pieces rip off easily if you tear between the slices; only Conan could rip off pieces horizontally.

Submitted by L. Stone, of Mamaroneck, New York. Thanks also to: Julia Berger, of Richmond, Virginia, and Virginia E. Griffin, of Salinas, California.

Why Are The Netherlands also Called Holland and the Low Countries? And Why Are Its People Called Dutch?

Our pet theory was that the official name of the country was "the Netherlands," but that "Holland" was used to make it easier for mapmakers to fit the name within the confined borders. Actually, the official name of the country is Nederland, the name native inhabitants call it—"Netherlands" is simply the closest English equivalent.

The word "nether" means below the earth's surface. The low and marshy lands near the mouth of the estuary of the Rhine River are responsible for the name, "the Low Countries." The German name "Niederlande" and the French name "les Pays-Bas" are exact translations.

By why "Holland"? Holland was the name of a province, not the whole country. In the sixteenth, seventeenth, and eighteenth centuries, it was by far the most important province commercially, and Hollanders displayed more devotion to their province than to the nation as a whole. Holland eventually became so dominant that, much in the same way that the Soviet Union is mislabeled "Russia," Holland came to represent all of the Netherlands.

Further confusing the issue is the term "Dutch," used to describe the citizens of the Netherlands. "Dutch" is actually older than "the Netherlands." Until the sixteenth century, inhabitants of the Netherlands called themselves Diets (which means "the people"). This word, pronounced "deets," was corrupted in English as "Dutch." The British continued to use the medieval name long after Netherlanders stopped using it themselves.

Americans tend to use the word "Dutch" not only to describe Netherlanders, but also Germans. Thus, while the Holland Dutch from Michigan are true descendants of Netherlanders, the Pennsylvania Dutch are actually German.

(The "Dutch" in Pennsylvania Dutch almost certainly stems from a corruption of the German name for their country, Deutschland.) According to the Netherlands Chamber of Commerce: "To stop this confusing multiplicity of names the Netherlands Government has tried to use the words 'The Netherlands' as the name for the country and 'Netherlander' as the name for an inhabitant of the Netherlands. It is easy to decree such a thing, but it takes much time to suppress a time-honoured word used in foreign countries."

Netherlanders have to deal with confusion not only about the name of their country, but about the name of their capital. Amsterdam is the official capital, but the seat of government is at The Hague. The official name of The Hague is 's-Gravenhage, "the count's hedge," except nobody calls the city 's-Gravenhage, preferring the colloquial Den Haag (the hedge).

For such a small country, the Netherlands has its share of identity problems.

Submitted by Daniel Marcus, of Watertown, Massachusetts.

What Are Those Twitches and Jerks That Occasionally Wake Us Just as We Are Falling Asleep?

It has probably happened to you. You are nestled snugly under the covers. You aren't quite asleep but you're not quite awake. Just as your brain waves start to slow, and as you fantasize about owning that Mercedes Benz convertible, you are jolted awake by an unaccountable spasm, usually in a leg.

You have been a victim of what is called a "hypnic jerk," a phenomenon explained in David Bodanis's marvelous *The Body Book*:

> They occur when nerve fibers leading to the leg, in a bundle nearly as thick as a pencil, suddenly fire in unison. Each tiny nerve in the bundle produces a harsh tightening of a tiny portion of muscle fiber that is linked to it down in the leg, and when they all fire together the leg twitches as a whole.

Sleep specialists haven't pinned down what causes hypnic jerks or why they occur only at the onset of sleep. Although some people experience them more often than others, their appearance is unpredictable, unlike myoclonic jerks, spasms that occur at regular intervals during deep sleep.

Submitted by Cathy C. Bodell, of Fullerton, California. Thanks also to: Daniel A. Placko, Jr., of Chicago, Illinois.

Why Are There Twenty-one Guns in a Twenty-one–Gun Salute?

The original intention of gun salutes was probably to assure the royalty or nation being honored that they were physically secure —that the weapons that were meant to pay tribute could also be used to kill. Before any recorded history of formal gun salutes, many cultures were known to discharge ordnance indiscriminately at festivals and holidays. Some good old-fashioned noise, be it fireworks in China or cheering at football games, has always been an accompaniment to joyous rituals.

Twenty-one–gun salutes have existed since at least the sixteenth century (the final scene from *Hamlet* mentions one), but the number of guns fired evolved gradually and inconsistently from country to country. The English were the first to codify the practice. According to a study conducted in 1890 by C. H. Davis, a commander and chief intelligence officer in the U.S. Navy, the earliest English regulation, formulated in 1688, prescribed that the birthdays and coronations of royalty should be solemnized "by the Fleet, Squadrons, and every single ship of war, by the discharge of such number of their great guns," but allowed that the number of guns used should be decided upon by the chief officer. By 1730, the British Naval Regulations were amended so that the number of guns discharged was still at the discretion of the chief officer, but was not to exceed twenty-one for each ship.

The notion of twenty-one as the highest gun salute undoubtedly stems from this royal origin. Salutes were always in odd numbers in the British military, with lower-ranking officers receiving, say, a five-gun salute and each increasing rank offered two more guns. The 1730 regulation was probably a response to rampant inflation in gun salutes; the Navy wanted to assure that no one received more guns than the royalty. In 1808, twenty-one guns was mandated as the *only* proper salute for royalty.

Although the United States, in its infancy, adopted gun salutes, there were no specific regulations governing the practice. Until 1841, the U.S. Navy fired one gun for each state in the Union. As our nation grew, and what with the price of ammunition, we prudently decided to limit our salutes to twenty-one guns. This change was codified in 1865 and has remained the practice ever since. The establishment of a maximum standard was not arbitrary or capricious. Gun salutes were a form of international diplomacy, and any deviation from the norm had possible ramifications. Commander Davis stressed the importance of conforming to international practice: "According to the present regulations and long established custom, a vessel of war, on her arrival in a foreign port, salutes the flag of the nation to which that port belongs, after having ascertained that the salute will be returned, with 21 guns. The salute is immediately returned gun for gun. This rule is universal and invariable in all countries in the world."

Davis believed that if the United States had continued its

practice of discharging one gun for each state, it could have had serious consequences. He feared that other countries would assume that by exceeding twenty-one guns, we were trumpeting our own power and superiority.

A last objection to the one gun/one state idea was that gun salutes have always involved an odd number. Even numbers have traditionally been fired in mourning and at funerals. While modern communications equipment has largely obviated the need to use gun salutes as a symbol of peace and goodwill, the twenty-one–gun salute is alive and well as a ritual to express celebration and honor. Although we can't pinpoint exactly why the British first arrived at twenty-one (some speculate that the combination of three multiplied by seven might have been adopted for mystical or religious reasons), we clearly owe our custom to the British military's desire to salute royalty with the utmost hoopla.

Submitted by Debra Kalkwarf, of Columbus, Indiana. Thanks also to: Douglas Watkins, Jr., of Hayward, California.

Why Do Women Tend to Have Higher Voices Than Men? Why Do Short People Tend to Have Higher Voices Than Tall People?

Daniel Boone, a University of Arizona professor and expert on vocal mechanisms, provides the answer: "Fundamental frequency or voice pitch level is directly related to the length and thickness of the individual's vocal folds [or vocal cords]." The average man's vocal-fold length is approximately eighteen millimeters; the average woman's is ten millimeters.

The tall person of either gender is likely to have longer vocal cords than a shorter person of the same sex.

Washington, D.C. Streets Are Named
Alphabetically. Why Is There No "J" Street?

We posed this Imponderable to Nelson Rimensnyder, historian of the House of Representatives Committee on the District of Columbia. Although Rimensnyder stated that there was no definitive answer, he did offer two main theories:

1. *J*, as written during the eighteenth century, was often confused with other letters of the alphabet, particularly *I*.

2. Pierre L'Enfant and other founders of Washington, D.C., were political, professional, and personal enemies of John Jay and therefore snubbed him when naming the streets in 1791.

Rimensnyder adds that there *is* a two-mile-long "Jay" Street in the Deanwood section of northeast Washington. Although this street presumably honors our first Supreme Court Chief Justice, its naming didn't upset Pierre L'Enfant in the slightest: "Jay" Street wasn't adopted until after 1900.

Submitted by M. Babe Penalver, of Bronx, New York.

What Happens to the Tread That Wears Off Tires?

The tread wears gradually off our tires. After a few years of heavy mileage, it eventually becomes bare. But we don't see bits of tread on the road (except from premature blowouts, of course). Highways are not discolored with blackened tread bits. Does tire tread disappear along with our socks?

The automobile industry, the tire industry, and some independent pollution experts have long been concerned about what may seem to be a trivial problem. Two specialists in the chemistry department of the Ford Motor Company have estimated that 600,000 metric tons of tire tread are worn off American vehicles every year. The possibility was more than remote that all of this material might remain in the air, in suspendable particles, which could be dangerous to humans. So they sought a way to measure what happens to the disappearing tread.

Tests to determine the presence of tire tread were held in three different sites, all of which presented some problems. First, indoor tests were designed to simulate driving wear on a tire. Unfortunately, without ambient weather conditions, worn rubber simply tended to stick to the simulated road surface. Scientists knew this wasn't what happened under real conditions, for the second type of tests, on real highways, indicated that virtually no rubber stayed on the road, due to wind, rain, and movement of surrounding traffic. Additionally, surface areas around highways were sometimes cleaned by maintenance crews, hindering efforts to measure long-term accumulation of tire tread. The third type of test, in tunnels, might be thought to show the maximum possible buildup of tire tread, except that road surfaces in tunnels tend to wear tires less than surface streets, and the lack of natural wind and rain in the tunnel made any extrapolation difficult.

Still, the combined results of these experiments did provide quite a lot of information about exactly what happens to tire tread. Whereas the most common substance in exhaust fumes is dangerous lead, the most plentiful tire debris is in the form of styrene-butadiene rubber (SBR), the most common rubber hydrocarbon in treads. Most of the tread debris is not in the form of gas, but rather in microscopic particles that are heavy enough to fall to the ground.

All road and tunnel tests seem to confirm that particle debris found along roadsides accounted for at least 50 percent of the total missing tire tread, and possibly much more. One study indicated that 2 percent of all roadside dustfall consisted of worn tread material. Another study, in Detroit, found that of the total particulate loading in the air, only 1 percent was tread dust. Even in tunnel tests, tire tread comprised only 1 to 4 percent of the total airborne particulate matter generated—a percentage far less than that of the exhaust emissions of gasoline- and diesel-powered vehicles.

All the tests concur, then, that the vast majority of worn tread in particle form falls on the ground instead of staying in the air.

What happens to the rest of the worn tread? Much of it is dissolved through oxidation and devulcanization (a chemical reaction that reverses the process used to harden rubber). One estimate speculated that devulcanization accounted for 30 percent of the disappearing SBR. Wind, water runoff, oxygen, and microbial attack all act to help degrade tread particulates, which degenerate faster than the tread rubber on tires in any case.

In fact, nobody could get very excited about the possible environmental dangers of worn tire tread. If the tread particulate were light enough to remain airborne, it could cause some harm, but the 95 percent plus that settles into the ground near the roadway poses no health hazard. K. L. Campbell, of Firestone Tire & Rubber Company, points out that "Tire tread rubber is essentially an inert material so it doesn't contribute to acid rain or soil pollution." And because worn tire-tread particles on the ground are in too small a form even to see with the naked eye, we aren't even aware that they are there. Which proves again that what you can't see can't hurt you.

Submitted by Larry Orbin, of Florissant, Missouri. Thanks also to: Brad Miles, of Victoria, British Columbia; G. William Foster, Jr., of Tulsa, Oklahoma; and Art Lombard, of Oakland, California.

Why Do Whips Make a Cracking Sound When Snapped?

Whips can attain a speed of more than seven hundred miles per hour when snapped, breaking the sound barrier. What you are hearing is a mini sonic boom.

HOW Did Xmas Come to Stand for Christmas?

The use of the colloquial "Xmas" has often been singled out as an example of how the holiday has been commercialized and robbed of its religious content. The *X* in Xmas is actually the descendant of the Greek equivalent of *Ch*, as in "Christos," which means "Christ." The letter *X* has stood for Christ (look up *X* in any dictionary) since at least A.D. 1100, and the term "Xmas" was first cited in 1551. Word expert Eric Partridge points out that the scholarly abbreviation for "Christianity" is "Xianity."

So many people dislike "Xmas" for its supposed crassness that its use is now virtually confined to commercial literature and banners. The *New York Times Manual of Style and Usage*, for example, offers this simple recommendation for when "Xmas" is acceptable: "Never use."

Submitted by Bobby Dalton, of Maryland Heights, Missouri. Thanks also to: Andrew Neiman, of Dallas, Texas.

Do Batteries Wear Out Faster If You Turn Up the Volume of a Radio?

Absolutely.

The battery applications manager of Eveready Battery Company, Inc., B. G. Merritt, told us about some research that proved the point conclusively:

> We recently tested a major manufacturer dual cassette "boom box" powered by 6 "D" size cells. From lowest setting to highest setting on the volume control, the power necessary to drive the "box" increased three times. This power increase directly translates into one third battery life at full volume when compared with zero volume. This power increase is necessary to drive the speakers.
>
> By comparison, a personal stereo (portable type) cassette player current increased only 30% when adjusted from zero volume to full volume. Battery life would be decreased only 30% for this device.

Don French, a battery expert at Radio Shack, confirmed Eveready's findings. He estimated that a shirt-pocket portable radio would use at least 200 percent more battery charge at the loudest volume setting than at the softest. French pointed out that even tiny radios have audio amplifiers that must be powered. A home stereo might require fifty watts and a shirt-pocket radio two hundred milliwatts, but the principle is the same—the more power required, the more juice required.

Submitted by Allen Kahn, of New York, New York.

Why Do Some Ranchers Hang Old Boots on Fenceposts?

It all started with an innocent call from a listener to Tannah Hirsch's KMBZ talk show in Kansas City. The listener asked the Imponderable above, and we admitted that we didn't have the slightest idea why some ranchers hang old boots on fenceposts. In fact, being urban folks, we had never seen this phenomenon at all. But it was an intriguing question, and we decided to investigate. The results will show you some of the trials and tribulations of tracking down answers to Imponderables.

A few weeks later, we received a letter from Rick Miller, who works at Kansas State University as an agricultural agent in Johnson County, Kansas. Rick had previously worked in southwestern Kansas, had seen boots hanging on fences, and had investigated. His conclusion was intriguing: "The boots are hung on the fence to discourage predators such as coyotes from entering fields where livestock are. The coyotes, with their keen sense of smell, pick up the human odor from the boots. Thinking humans are around, the coyotes won't cross the fence." Having read *All the President's Men* and faithfully watched *Lou Grant*, we were determined to dig up a second source to corroborate Mr. Miller's story.

First we wrote to several ranchers' organizations. All were familiar with the practice, but none mentioned the coyote angle, and only one had any theory at all—Rhoda G. Cook, executive secretary of the Montana Outfitters and Guides Association: "My late husband, who was a long-time packer and breaker of mules for the Forest Service said several times his boots were on the fence so everybody would know that a real cowboy lived there. I can vouch for the fact that they certainly smelled better out there than under the bed."

Still missing the "smoking gun" that could definitively answer the Imponderable, we contacted all of the largest manufacturers of cowboy boots. Nobody could help us, but Frye sent us to *Western Horseman* magazine, where we spoke to writer Darrell Arnold and publisher Dick Spencer. Mr. Spencer didn't put much stock in the coyote theory, but he remembered seeing long stretches of boots on fenceposts along Windless Hills, near Ogallala, Nebraska. He recommended that we contact someone in that area.

We found out that the newspaper of record in Ogallala was the Keith County *News*, published by Jack Pollack. Mr. Pollack was quite familiar with the stretches of boots on fenceposts, but didn't know why they were there. He suggested we speak to some ranchers in the area and was kind enough to supply some phone numbers.

We contacted several ranchers, none of whom could say how the practice began. But one rancher, Waldo Haythorne, asked if we wanted to get in touch with the son of the man who started the tradition. We sure did!

And we found out that the pioneer in the practice of hanging boots from fenceposts was none other than Henry Swanson. According to his son, Virgil Swanson, after Henry put some worn boots on his fenceposts his neighbors followed his example, until the path from the town of Arthur to the cemetery about a quarter-mile away was marked by boots at every ranch. Others copied the practice, and up in the hills above Arthur, there are longer stretches of boot-strewn fences (probably the ones Dick Spencer had spotted).

So, with great anticipation, we asked Virgil Swanson the question that had come to obsess us: WHY did your father hang the boots from the fencepost? He paused for a moment, and then replied, "Gee. I really don't know. I guess he just did it to do it."

Dejected, we were back to first base. But quitters we are not. We called up our original correspondent, Rick Miller, to find out more about the coyote angle. Rick told us that coyotes have a keen sense of smell, and are uncommonly smart animals. Coyotes don't like to mess around with people, and boots trapped the human scent better than anything else. Rick said that some ranchers in southwestern Kansas put boots on every single fencepost surrounding their fields, and it seemed to work for them. We were still a tad skeptical, because we couldn't get anyone else, including a few vets we consulted, to confirm Rick Miller's explanation. Rick suggested we speak to Bob Henderson, who works for the Kansas State Wildlife Department.

Bob Henderson, of course, did *not* confirm Rick's theory. He felt that coyotes are too smart to be fooled more than once or twice by boots on fenceposts, and that coyotes are not as frightened by human odors as most people think. The problem with the boot-as-repellent theory is that coyotes, despite their excellent sense of smell, do not rely only on that sense. The coyote also uses its sight to determine whether it is safe to prey; once it becomes accustomed to seeing the same boots on the same fenceposts day after day, it won't be timid about scaling the fence. In fact, Henderson said, coyotes are not afraid of human scent (human urine has even been used as an attractant to trap coyotes)—they are afraid of people.

Just as we were about ready to give up on this Imponderable

(or put it in the Frustrables section), we received a letter from Lynda Frank, of Omaha, Nebraska, posing the same Imponderable. We called her immediately. Although Lynda had no concrete theories, she assumed that the practice was simply a traditional one, without any practical purpose.

We then spoke to five or six experts in Animal Damage Control. These are local or regional governmental agencies that attempt to rid areas of coyotes and other pests. None of them was familiar with the boot-repellent theory. One person referred us to Dr. Mike Fall, at the Denver Wildlife Research Center, a specialist in predator control. Like most good scientists, Dr. Fall was circumspect about speculating on a topic he had not researched himself, but he had grave doubts about whether boots on fenceposts would have any long-lasting effect on deterring coyotes from trespassing.

Dr. Fall emphasized that to discourage coyotes, you must chip away at one of their strengths—their adaptability. You must disrupt the patterns they encounter, for they can become accustomed to just about anything, including human odors. Scientists still don't fully understand the impact of odors on coyotes. Scents have been developed that can consistently attract coyotes, but nothing yet can consistently repel them.

Fall predicted that boots on fenceposts would work over time only if the footwear were constantly rearranged, making the coyote insecure about whether humans were afoot. Experiments have been conducted using electric fences, guard dogs, and other animals, lights, and loud radios and tape recordings to deter coyotes, all with some, but limited success. The most promising approach, according to Dr. Fall, is a multistimulus deterrent, flashing lights with high-frequency sounds produced in *random patterns,* so that the coyotes are thrown off guard.

So—after consulting nearly thirty people, we still don't have a definitive answer to this Imponderable. It has thoroughly humbled us. We would love to hear from readers who might have the answer. Until then, we can only offer the three theories offered to us that make sense:

1. The boots scare away coyotes.

2. It is far wiser to stink up a fencepost than to stink up a house.

3. To paraphrase Virgil Swanson, "Some people just do things to do things."

Submitted by Rick Miller, of Gardner, Kansas. Thanks also to: Lynda Frank, of Omaha, Nebraska.

Why Do Bananas, Unlike Other Fruits, Grow Upward?

If you knew about the tumultuous birth process of the banana, perhaps you would be more charitable the next time you encounter some bruised specimens at the supermarket.

The banana is actually a giant herb in the same biological family as lilies, orchids, and palms. It is the largest plant on earth without a woody stem—a banana stalk is 93 percent water—and is consequently extremely fragile. Although it can reach a full height of fifteen to thirty feet in one year, even moderate winds can blow down a plant.

The fruit stem or bunch originates at ground level. At this stage, the bunch consists of all of the fruit enclosed in leaf bracts. The individual fruit "fingers" (the technical name for a single banana) are pointed upward. As the bunch or bud is pushing its way through the mass of tightly packed leaf sheaths known as the pseudostem, the fruit fingers remain pointed upward until they emerge at the top of the plant.

The bananas exert tremendous pressure on the pseudostem. Before the fruits expand, the leaves enclosing them roll around on themselves inside the trunk. After the fruit emerges from the leaves, the fingers point downward, but only because the bud surrounding them has changed direction.

Once the entire bunch of bananas is mature, fully emerged from its sheath, and pointing downward, the individual leaf bracts enclosing the hands (the female flower cluster) fall away, exposing the fruit. At this point, the individual flowers grow rapidly, filling out. Their increased weight bends the main stalk so that the individual fruits on the hand start to turn upward in about seven to ten days.

Dr. Pedro Sole, of Chiquita Brands, points out that in the past, "primitive bananas grew upwards, like the seeds of most grasses forming a spike."

So is there a logical reason for the banana's tortuous up-and-down birthing process now? Jack D. DeMent, of the Dole Fresh Fruit Company, sees the answer in the behavior of the traditional noncommercial banana plant:

> A flower is found on the tip of each individual fruit. This flower is removed during [commercial] packing but is present during fruit development. As the hands turn up, the flower is better exposed to insects and nectar eating birds and bats. Their feeding would normally aid in fertilization of the fruit. Today's commercial banana is sterile and rarely—almost never—produces a viable seed.

DeMent theorizes that the commercial banana's tendency to grow upward is a holdover from its ancestors that needed to point upward for their very survival.

Presumably, natural selection will simplify the growth process of the banana over the next few hundred thousand years or so.

Submitted by Lynda J. Turner, of Hackettstown, New Jersey.

Why Is There a Black Dot in the Middle of Otherwise White Bird Droppings?

An important question, one that philosophers throughout the ages have pondered. Luckily, ornithologists know the answer.

That black dot is fecal matter. The white stuff is urine. The urine and fecal matter of birds collect together and are voided simultaneously out of the same orifice. Feces tend to sit directly in the middle of droppings because the urine, slightly sticky in consistency, clings to them.

Submitted by Ann Marie Byrne, of Queens, New York.

Do Toilet-Seat Covers Really Protect Us Against Anything?

We became suspicious about the efficacy of toilet-seat covers when we pondered, one day: why don't they sell toilet-seat covers for home use? You see them only in public rest rooms. Perhaps the idea is not to protect you from disease, but from the thought of exposing your bare backside to the same surface area occupied by heaven knows who before you.

We were on the right track. Not only are venereal diseases *not* spread by toilet seats, but nothing else is, either. Although there was one report suggesting that the herpes virus *may* survive *briefly* in such an environment, the secretary of the American Society of Colon and Rectal Surgeons, Dr. J. Byron Gathright, Jr., echoed the sentiments of other doctors we spoke to: "There is no scientific evidence of disease transmission from toilet seats."

Submitted by Jean Hanamoto, of Morgan Hill, California.

Why Do Sailors Wear Bell-Bottom Trousers?

Nobody knows for sure if there was one particular reason why this custom started, but three theories predominate:

1. The flared leg allows bell-bottoms to fit over boots easily. Sailors traditionally sleep with their boots at the side of the bed, so that, in case of emergency, they don't have to waste time trying to position their pants over their footwear. Once a sailor arrives on deck, having the trouser legs fully cover the top of the boot has practical advantages as well—it protects him from spray and rain entering his boots.

2. Bell-bottoms are easily rolled up. Because sailors often work with potentially harmful chemicals (scrubbing the deck with lye, for example), rolling up the cuffs prevents permanent damage to the pants. Also, if a sailor needs to wade ashore, bell-bottoms can easily be rolled up above the knee.

3. If a sailor is thrown overboard, bell-bottoms are also easier to remove than conventional trousers. And the loose fit of the
bell-bottom also makes it easier to remove boots in the water.

Sailors in boot camp are taught another practical use for bell-
bottom trousers. If the legs are tied at the ends, bell-bottoms can
hold quite a lot of air; in a pinch, they can be used as flotation
devices.

Why Doesn't Sugar Spoil or Get Moldy?

Virtually all living organisms can digest sugar easily. So why
isn't sugar prone to the same infestation as flour or other kitchen
staples?

Because sugar has an extremely low moisture content—usually about 0.02 percent—it dehydrates microorganisms that
might cause mold. As John A. Kolberg, vice-president of opera-
tions at the Spreckels Division of Amstar Corporation, explains
it, "Water molecules diffuse or migrate out of the microorganism
at a faster rate than they diffuse into it. Thus, eventually the
microorganism dies due to a lack of moisture within it." Sugar's
low moisture level also impedes chemical changes that could
cause spoilage.

All bets are off, however, if sugar is dissolved in water. The
more dilute the sugar solution, the more likely yeasts and molds
will thrive in it. Even exposure to high humidity for a few days
will allow sugar to absorb enough moisture to promote spoilage
and mold.

Storing sugar in an airtight container will retard the absorp-
tion of moisture even in humid conditions. If stored in an atmosphere unaffected by swings in temperature and humidity, sugar
retains its 0.02 percent moisture level and has an unlimited shelf
life.

Submitted by Joel Kuni, of Kirkland, Washington.

Why Do Nurses Wear White? Why Do Surgeons Wear Blue or Green When Operating?

Florence Nightingale always wore a white uniform. White, of course, is a symbol of purity, and in the case of a nurse, an appropriate and practical one—white quickly shows any dirtiness.

Surgeons also wore white until 1914, when a surgeon decided that red blood against a white uniform was rather repulsive and needlessly graphic. The spinach green color he chose to replace it helped neutralize the bright red.

At the end of World War II, the lighting was changed in operating rooms, and most surgeons switched to a color called "misty green." Since about 1960, most surgeons have used a color called "seal blue," which contains a lot of gray. Why this latest switch? According to Bernard Lepper, of the Career Apparel Institute of New York City, seal blue shows up better on the TV monitors used to demonstrate surgical techniques to medical students.

Submitted by Norman J. Sanchez, of Baton Rouge, Louisiana. Thanks also to: Lori Bending, of Des Plaines, Illinois; Andrew Neiman, of Dallas, Texas; and Reverend Ken Vogler, of Jeffersonville, Indiana.

Why Doesn't a "Two-by-Four" Measure Two Inches by Four Inches?

Before the invention of mass-scale surfacing equipment, most lumber was sold to the construction trade in rough form. In the "good old days," a "two-by-four" was approximately two inches by four inches. Even then, two inches by four inches was a rough estimate—cutting equipment trimmed too thick or too thin on occasion.

As the construction trade demanded smooth edges, surfacing machinery was created to handle the task automatically. These devices reduced the dimensions of the rough lumber by at least one-eighth of an inch in thickness and width.

The radio talk-show caller who posed this Imponderable wondered why he got gypped by buying finished "two-by-fours" that measured 1⅝ inches thick by 3⅝ inches wide. The answer comes from H. M. Niebling, executive vice-president of the North American Wholesale Lumber Association, Inc.:

> [After the early planers were used,] profile or "splitter" heads were developed for planers, wherein one could take a 2″ × 12″ rough piece and make 3 pieces of 2 × 4s in one surfacing operation [i.e., as the lumber went through the planer it was surfaced on four sides and then, at the end of the machine, split and surfaced on the interior sides]. Unfortunately, the "kerf," or amount of wood taken out in this splitting operation, further reduced the widths.

The size of these "kerfs," three-eighths of an inch, didn't allow processors to make three pieces 3⅞ inches wide (three times 3⅞ plus three times ⅜, to represent the "wastage" of the kerfs, equals 12¾ inches, wider than the original 12-inch rough piece). This is why the dimensions of the finished piece were reduced to 1⅝ inches thick by 3⅝ inches wide.

If you think this is complicated, Niebling recounts other problems in settling the dimensions of lumber. Fresh-cut lumber is called "green" lumber, whether or not it is actually green in color at the time. Green lumber must be dried by natural or artificial means. When lumber dries, it shrinks and becomes stronger. Some lumbermen believed that either dry lumber should be sold smaller in size or that green lumber should be sold larger. Recounts Niebling: "The result was that 2 × 4s surfaced dry comes out at 1½″ by 3½″ instead of 1⅝″ by 3⅞″. To settle the fight between green and dry producers, a green 2 × 4 is surfaced to 1⁹⁄₁₆″ by 3⁹⁄₁₆″. In effect, they reduced the green size too to settle the fight."

The lumbermen we spoke to agreed that the pint-sized two-by-fours provided the same strong foundation for houses that the rough original-sized ones would. One expert compared the purchase of a two-by-four to buying a steak. You buy a nice steak and it is trimmed with fat. Sure, the butcher will trim off the fat, but then he'll raise the price per pound. One way or the other, you pay.

Why Is an Acre 43,560 Square Feet?

"Acre" is an Anglo-Saxon word that means, literally, the amount of land plowable in one day. The term was used before the tenth century, the acre originally referring to the area that could be plowed by a yoke of oxen in one day. The actual footage of the acre varied from region to region.

In the late thirteenth and early fourteenth centuries, Edward I and Edward III tried to codify English measurements. Although the quantity of land that could be plowed in one day was obviously variable, depending upon such factors as the durability of the animals pulling the plow, the plowing equipment, and the topography of the land, there were obvious advantages to standardization. By the reign of Henry VIII, there was universal agreement that an acre should be 40 poles long by 4 poles wide (or 160 square rods). These nice round units of measurement (one rod = 16.5 feet; one pole = one square rod), popular in agricultural societies, translate exactly to our current standard of the acre as 43,560 square feet. With modern machinery, any farmer can plow considerably more than one acre in a day, but the acre has proved to be an enduring unit of measurement.

Why Do Men's Bicycles Have a Crossbar?

We're sure you'll be overjoyed to learn that everyone we talked to agreed on the paramount issue: that crossbar at the top of the frame makes men's bikes far sturdier than women's. After centuries of experimentation, manufacturers have found that the best strength-to-weight ratio is maintained by building frames in the shape of diamonds or triangles. Without the crossbar, or as it is now called, the "top tube," part of the ideal diamond structure is missing.

A man's bicycle has its top tube parallel to the ground; on a ladies' bicycle, the top tube intersects the seat tube several inches above the crank axle. Why is the women's top tube lower than the male's?

The tradition is there for no other reason than to protect the dignity and reputations of women riding a bicycle while wearing a skirt or dress. Now that most women bicyclists wear pants or fancy bicycle tights, the original purpose for the crossbar is moot, although Joe Skrivan, a product-development engineer for Huffy, points out an additional bonus of the lower top tube: it allows for easy mounting and dismounting.

Skrivan notes that the design difference creates few complaints from women. Casual women bicyclists don't necessarily need the rigidity of the higher crossbar. Serious female bicyclists buy frames with exactly the same design as men's.

Submitted by Linda Jackson, of Buffalo, New York.

Why Is Royalty Referred to as "Blue-Blooded"?

In the eighth century, a group of Islamic warriors, the Moors, invaded and occupied Spain. And they ruled over the country for five centuries.

This didn't sit too well with the aristocrats of Castile, who began referring to themselves as *sangre azul* ("blue blood") to differentiate themselves from the Moors. No, the Castilians' blood was no different in color than the Moors, but their skin complexion was lighter than their conquerors.

The Castilian pride in their "blue blood" was a thinly veiled proclamation of pride in their light complexions, and a subtle way of indicating that they were not, as the *Oxford English Dictionary* puts it, "contaminated by Moorish, Jewish, or other foreign admixture." For the paler the complexion of the skin, the more blue the veins appear.

Submitted by Daniel A. Placko, of Chicago, Illinois.

Why Are People Immune to Their Own Body Odor?

How can so many otherwise sensitive people expose others to their body odors? Surely, they must not know that they (or their clothes) are foul-smelling, or they would do something about it. Right?

Right. Compared to most animals, humans don't have an acutely developed sense of smell. According to Dr. Pat Barelli, secretary of the American Rhinologic Society, "The olfactory nerve easily becomes 'fatigued' in areas where there are odors." In order not to be overloaded with information, your nervous system decides not to even try being "bothered" by your body odor unless it changes dramatically. Whether you regularly smell like a spring bouquet or like last night's table scraps, you are unlikely to notice—even if you are sensitive to the body odor of other people.

Dr. Morley Kare, director of the Monell Institute at the University of Pennsylvania, adds that this fatigue principle applies to many of the senses. Workers at automobile factories must learn to block out the sounds of machinery or risk being driven insane. Residents of Hershey, Pennsylvania, stop noticing the smell of chocolate that permeates the town.

Students often can't discriminate the taste of different dishes served in their school cafeteria. Of course, this phenomenon might be explained by the fact that all the cafeteria dishes *do* taste alike, but we would need a government grant to confirm the thesis.

Submitted by Karole Rathouz, of Mehlville, Missouri.

Why Are the Outside Edges of the Pages of Many Paperback Books Colored?

In the early days of paperback books, the paper used was of very low quality, usually newsprint. Consumers rejected the soiled and discolored appearance of the pages. Publishers hit upon the notion of "staining," which made the paper look fresh, even pretty, and most important, prolonged the shelf life of their books.

Some publishers used the same color stain for long periods of time, in an attempt to make their company's product easily identifiable in the bookstore. For a long time, Dell's paperbacks were stained blue; Bantam's were yellow; Pocket Books favored red.

As the paper quality improved, the necessity for staining decreased. Some publishers still stain some of their mass-market (small-sized) paperback books. Occasionally, even today, the paper quality is low, or the paper within one book varies slightly in color—staining eliminates these problems. Trade (larger-sized) paperbacks use higher-quality paper, so staining is rare. Ironically, the tradition of staining dates back to the days of Gutenberg, when Bibles were stained for aesthetic purposes. Some expensive hardcover books are stained today to add a touch of panache.

Paperback books are stained by machine after they are completely bound. The books are moved on a conveyor belt that has sides and walls to protect the books from errant ink. Two jets spray ink all over the top, bottom, and nonbound side of the paper.

The staining of hardbound books used to be done by machine, but since the practice has almost completely died out the machinery has been sold off. Today, staining of hardbounds is done by hand, with a spray gun. The books are taken off the assembly line before they are cased. Protected by backboards and wings, the books are sprayed three at a time. The ink dries exceptionally fast.

Although staining adds some expense to the production cost, publishers must wonder: Does anyone notice? Does anybody care? The production experts we spoke to felt that the custom of staining persisted more because of inertia than for any practical purpose.

Submitted by Pat O'Conner, of Brooklyn, New York.

How Do They Shell Pine Nuts?

With great difficulty.

Paul Wallach, who hosts a popular interview show in Los Angeles concerned largely with food and restaurants, told us that this Imponderable had stumped him for a long time. What machine, he wondered, could possibly be fitted to work on pine nuts?

It turns out that no machine works consistently well in shelling pine nuts. Most pine-nut processors use almond shellers, which do only a decent job of shelling without ruining the nutmeat.

Many of the pine nuts from China are shelled by hand. Or rather, by a hammer held by a human hand. Not high-tech. Not fast. Labor intensive. But effective.

Submitted by Paul Wallach, of Los Angeles, California.

How Can Owners of Small Cemeteries Make Money? How Can They Plan Their Finances When They Have to Wait for People to Die Before They Derive Income?

We were asked this Imponderable several times on radio talk shows. And we were stumped. The income of a small cemetery owner must be severely limited by the population the cemetery serves. In many cases, privately owned cemeteries and funeral homes even in the smallest towns must "compete" against their church-owned or municipal counterparts. Church-owned cemeteries often charge only for the cost of digging a grave; the privately owned cemetery charges Tiffany prices in comparison.

We found out that more than a few cemetery owners in small towns are not millionaires. Many funeral directors and a few cemeterians need second jobs to provide more income. How do the small cemeterians survive? Are there any (legal) ways of "drumming up" business?

We were lucky enough to find Howard Fletcher, the chairman of the Small Cemetery Advisory Committee of the American Cemetery Association. Mr. Fletcher, who owns a memorial park in Muscatine, Iowa, helps fellow small cemeterians contend with the very financial problems we have discussed. Despite all the jokes about the business (such as "*everybody* is a potential customer"), a small cemeterian must do more than sit around and wait for people to die in order to survive. Howard Fletcher is unusually frank and unsanctimonious about his profession, and unashamed about the methods he uses to maximize his income. He developed a pamphlet called "50 Sources of Income for Small Cemeteries," from which most of the material below was adapted.

Within Fletcher's fifty sources of income are at least five broad categories: preselling; upgrading; maximizing underutilized assets; creative financing; and expanding services and products.

Preselling

To Fletcher, this is the key ingedient in a successful small cemetery operation. Most funeral directors have to wait until a death before seeing any income. Fletcher tries to sell his community on the advantages of buying space, vaults, caskets, and even memorial markers "preneed" rather than "at need." He has many arguments in his arsenal: a preneed purchase saves the bereaved family from the emotional strain of making funeral arrangements at the time a loss occurs; the decision can be made at the home of the buyer; prices will be lower now than when bought in the future; no cash is necessary right away, while most funeral directors would require some cash "at need"; making arrangements now will provide the buyer with peace of mind, not only for him or herself, but in knowing that the family will not be saddled with the unpleasant task; spouses can make decisions about funeral arrangements together; terms are negotiable—the buyer is likely to have more leverage when he or she is hale and hearty. To quote Mr. Fletcher: "It is not a question of if these arrangements will be made, it is only a question of who is going to make them and when!"

Here are some of the successful variations of preneed selling:

1. Sell child burial protection. Child protection doesn't cost much, but it does provide great cash flow. By the time the child is likely to die, compound interest has made this presell very profitable.

2. Presell grave opening and closing charges.

3. Offer one free burial space or two-for-one sales to married couples. Presumably, married couples want to be buried together, so the free space for one turns out to be the same deal as the two-for-one—these offers are always nonassignable and nontransferable (thus solving the possible divorce problem).

Upgrading

1. Sell marker refinishing kits. Bronze markers often tarnish because of oxidation.

2. Sell granite bases as upgrades from concrete bases.

3. Sell larger memorials.

4. "Reload." Use existing customers as a base to sell new or improved products. This is one reason cemeterians like to deliver by hand all deeds and official papers. They can discreetly get referrals or find family members who have not yet made funeral plans. Fletcher issues two newsletters per year with return cards and pitches for upgrading products.

5. Sell wreath and grave coverings for Christmas, Memorial Day, and other holidays.

6. Sell vesper lights.

7. Sell carillon chimes with the donor's name on plaque.

Maximizing Underutilized Assets

1. Launch a lawn-care business to more fully utilize landscaping equipment.
2. Sell double-depth privileges.
3. Grow and sell sod.
4. Raise and sell nursery stock from open land.
5. Cut and sell firewood from open land.
6. Sell excess trees on property.
7. Lease extra acreage to farmers.
8. Sell excess materials from graves as fill dirt.

Creative Financing

Many of these tips consist of charging separate fees for services that might or might not be included in the usual package deal:

1. Charge a filing and recording fee.
2. Charge for deed transfer and replacement.
3. Offer discount for cash payment of open accounts in order to generate cash flow.
4. Sell accounts receivable for cash flow.
5. Charge interest on house accounts.
6. Increase price of lots by having care charge paid separately.
7. Increase price of memorial by having installation and care charges paid separately.
8. Sell for allied businesses, such as monument dealers.
9. Sell extra-care charge for special care.
10. Where cemetery has historic value, apply for federal, state, or local registry in historical society for funding purposes.
11. Hire professional collectors for delinquent accounts.
12. Offer a discount on a new marker if purchased within one month of burial.

Expanding Services and Products

Here are some of the more creative ideas, all potentially practical:

1. Start a pet cemetery, with preneed and at need sales.
2. Manufacture vaults.
3. Build a funeral home that offers preneed as well as at need follow-up.
4. Start a trailer park on extra acreage where feasible.
5. Rent the chapel tent for weddings and lawn parties.
6. Raise and sell livestock.
7. Develop a flower shop.
8. Sell garden features and entrance features.
9. Sell trees—lining drives and/or walks.
10. Sell benches in cemetery.
11. Sell stained-glass windows.
12. Sell pews in chapel.
13. Sell furniture in mausoleum or committal area.

Some of these "money-making tips" might be offensive to your sensibilities. The image of a trailer park next to the memorial park is less than pleasing, and the thought of discussing preneed services at your kitchen table might dull the appetite a bit. The alternative, though, is usually a full-court press at the time of death.

Howard Fletcher is providing a service, but is also willing to admit that he is in business to make money. He wants the public to know what the business is like, so that the public can understand the industry's problems. Most small cemeteries make less than $100,000 in sales per year and conduct fewer than 150 burials. In order to survive, the small cemetery owner must often hustle as aggressively as any other salesperson.

Why Are Most Homes Painted White?

Most homes in the United States have always been painted white. Paint was first used as a preservative as much as an aesthetic expression. White was evidently believed to be more durable than other mixtures, but there were also historical reasons for its popularity. White was associated with the classic Greek and Roman architectural forms. Furthermore, Puritans viewed color as frivolous; the "seriousness" of white continued to appeal to Americans as late as the mid-nineteenth century.

In 1842, American architect Andrew Jackson Downing launched an attack against the color white for homes (a large proportion of American homes were then painted white with green shutters):

> There is one colour . . . frequently employed by house painters, which we feel bound to protest against most heartily, as entirely unsuitable, and in bad taste. This is white, which is so universally

applied to our wooden houses of every size and description. The glaring nature of this colour, when seen in contrast with the soft green foliage, renders it extremely unpleasant to an eye attuned to harmony of coloring, and nothing but its very great prevalence in the United States could render even men of some taste so heedless of its bad effect.

Downing argued for muted earth tones as the best alternative to white, and for a while his aesthetic was influential, especially after the paint industry developed the technology to premix paints of various shades and ship them safely throughout the country by rail. (Until after the Civil War, local painters had had to mix dry colors with lead and oil to create nonwhite shades of paint.)

In the late nineteenth century, white houses became the vogue once again, and although tastes in home colors have gone through many cycles in the past hundred years, white has never become unfashionable. A survey of paint authorities yielded some reasons for its endurance as our most popular color:

1. The choice of white can never be a disaster. Although you risk seeming unchic by avoiding a more "daring" color, you can never be accused of tackiness. Shari Hiller, the color stylist for Dutch Boy Paints, thinks this is the most important reason for the popularity of white:

I have found in putting together color cards for our brands, that the homeowner is pleading for suggestions in the exterior color scheme areas. When we finally answer our customers' needs and provide them with enough advertising, photos, and helpful suggestions that they feel more comfortable making a color decision, I think we may see many other colors gaining in popularity.

2. White has so many pleasant associations. White connotes cleanliness, peace, strength, and purity.

3. White is classic. Much like the basic black dress, white is unlikely ever to go out of style. Our president, after all, doesn't live in the Puce House. Of our nation's major monuments, only the Statue of Liberty isn't white (and the statue, of course, was a French import).

4. White goes well with other colors. White mixes well with any shutter trim the homeowner desires, and with all roof colors.

Submitted by Mark Carroll, of Nashville, Tennessee.

Why Is One Side of Reynolds Wrap Aluminum Foil Shiny and the Other Side Dull?

Grown people, though no personal friends of ours, have been known to argue about whether the shiny side of Reynolds Wrap is supposed to cover the food or to be the side exposed to the outside elements. According to the folks at Reynolds Metals, it makes little difference which side of Reynolds Wrap you use. There is a slight difference in the reflectivity of the two sides, but the difference is so small that it can only be measured by laboratory instruments. Nikki P. Martin, Reynolds's consumer services representative, puts it succinctly if self-servingly: "Both sides do the same fine job of keeping hot foods hot, cold foods cold, wet foods wet, dry foods dry and all foods fresh longer."

Foil starts as a large block of solid aluminum. The block is rolled like a pie crust until it becomes one long, thin, continuous sheet. The dissimilar finishes of Reynolds Wrap are the result rather than the intention of its manufacturing process. Martin explains that "In the final rolling step, two layers of aluminum foil are passed through the rolling mill at the same time. The side coming in contact with the mill's highly polished steel rollers becomes shiny. The other side, not coming in contact with the heavy roller, comes out with a matte finish."

Submitted by Frank Russell, of Columbia, Missouri.

Why Do Superficial Paper Cuts Tend to Hurt More Than Grosser Cuts?

Perhaps paper cuts hurt more because they are so emotionally maddening. How can such a trivial little cut, sometimes without a hint of blood, cause such pain?

The sensory nerve endings are located close to the skin surface, and the hands, where most paper cuts occur, contain more nerve endings than almost any other area of the body. Dr. John Cook, of the Georgia Dermatology and Skin Cancer Clinic, adds that a trivial laceration such as a paper cut creates the worst of both worlds: "It irritates these nerve endings but doesn't damage them very much." Damaged nerve endings can lead to more serious complications, but sometimes to less pain than paper cuts.

Dr. Cook and Dr. Elliot, of the American Dermatological

Association, also mentioned that most patients tend not to treat paper cuts as they would grosser ones. After any kind of cut, the skin starts drying and pulling apart, exposing nerve endings. Cuts are also exposed to foreign substances, such as soap, liquids, perspiration, and dirt. Putting a bandage over a paper cut will not make it heal faster, necessarily, but if the cut stays moist, it won't hurt as much.

Why Does the Brightest Setting of a Three-Way Light Bulb Always Burn Out First?

As we sit typing this in the light of a General Electric 50/200/250-watt three-way light bulb, having experienced this plight many a time in the past, we took a personal interest in solving this Imponderable. If you have read the following Imponderable (and shame on you if you are reading out of order), you have already figured out the answer. When you can no longer get the 250-watt light, the reason is that the 200-watt filament has burnt out. All that is left is the 50-watt filament, lovely for helping plants grow, but hardly sufficient illumination in which to create literary masterpieces.

However, the higher-wattage filament doesn't *necessarily* burn out first. It does have a shorter rated life than the low-wattage filament. General Electric's research has shown that because the lower filament is often used as a night light or background light, it tends to get more use than the higher-wattage filament, so it is intentionally designed to have a longer life.

Submitted by Tom O'Brien, of Los Angeles, California.

HOW Do Three-Way Light Bulbs Work? How Do the Bulbs "Know" at Which Intensity to Shine?

Each three-way light bulb contains two filaments. Let's take as an example the popular 50/100/150-watt three-way bulb. When you turn the switch to the first setting, the lower wattage (50-watt) filament lights. When you turn to the next setting, the 100-watt filament lights and the 50-watt filament turns off. When you turn the switch for the third time, both the 50- and the 100-watt filaments light. This explains why the highest wattage rating for a three-way bulb is always the sum of the two lower wattage figures.

James Jensen, of the General Electric Lighting Business Group, is quick to explain that three-ways will work only in sockets designed to accept this type of bulb. While the three-way bulb, like conventional bulbs, makes contact in the socket through its screw shell and through an eyelet contact at the bottom of the base, it also contains a third feature. Says Jensen: "In addition, there is a contact ring surrounding the eyelet. This ring contacts a small post contact in the socket. Sometimes, a three-way bulb will flicker or fail to light on all settings. This is often due to poor (or no) contact in the socket. Sometimes merely tightening the bulb in the socket will remedy this."

Submitted by Elaine Murray, of Los Gatos, California.

Why Do Snakes Dart Out Their Tongues?

Although snake watchers at zoos love to see the reptiles flick their tongues, imagining they are ready to pounce on some unsuspecting prey, the tongues are perfectly harmless. Snakes don't sting or use their forked tongues as weapons.

The tongue is actually an invaluable sensory organ for the snake. It enables the reptile to troll for food (just as a fisherman sticks his line out in the water and hopes for the best), while feeling its way over the ground. It does this by bringing in bits of organic matter that it can smell or taste, alerting it to a potential food source. Some evidence suggests that a snake's tongue is equally sensitive to sound vibrations, warning it of potential prey or predators.

Where Do They Get That Organ Music in Skating Rinks?

As we discussed in our first volume of Imponderables, skaters are not allowed to use music with vocals in competitions, and we explained some of the reasons why that music sounds so awful. The inevitable follow-up question: what about the music in ice and roller skating rinks?

Chances are very, very good that any organ music you hear in skating rinks comes from a company called Rinx Records, the only known source for tempo organ music. Competitive skaters need all-instrumental music of specific lengths (usually three or four minutes, exactly) for competitions and achievement tests. Not only do these songs need to be an exact length, but many need to be an exact number of beats per minute. Rinx Records, for example, provide waltzes with 108, 120, and 138 beats per minute. The records must have a strong beat so that skaters can synchronize their movements with music often piped through horrendous sound systems.

Rinx Records was founded in 1950, in Denver, Colorado, by Fred Bergen, a man who not only was involved in skating, but was an organist who played on many records. In 1968, Bergen sold Rinx Records to Dominic Cangelosi, who still operates the business from the roller rink he owns. Cangelosi has played keyboards on all of the records he has released since 1968. His music is heard throughout the world, but like the baseball stadium organist, he labors in semiobscurity, unmolested by rabid fans on the street.

Rinx is a nice business. Although a few other individuals besides Cangelosi market tapes, Cangelosi has the record end of the field sewn up. He has a big market, with a mailing list of more than five thousand customers, including not only rinks but skating instructors and individual skaters as well. Ice skating and roller skating share many of the same tempos (though some ice skating music is much faster), so Rinx sells to both markets. In all, Rinx has more than thirteen hundred *different* records in stock, on seven-inch 45 rpm. If your heart prompts, you can find out more about Rinx Records by contacting Dominic Cangelosi at: P. O. Box 6607, Burbank, CA 91510.

Although Rinx's variety of organ music is associated with bygone days, Cangelosi has tried to spice up his arrangements with synthesizers, pianos, and electronic and Hammond organs in addition to the traditional acoustic and pipe organs. On some records, he adds guitar, drums, or other accompaniment. Cangelosi also "covers" popular songs, for which he pays a fee to ASCAP or BMI. Rink operators likewise have to pay a nominal fee to these licensing organizations for playing contemporary songs in their rinks.

George Pickard, executive director of the Roller Skating Rink Operators Association, says that most rinks have abandoned old-fashioned music for rock and disco. But many have special adult sessions that use Rinx and other more traditional records. There are even a few rinks that still have live organ music, the last echo of bygone days.

Submitted by Gail Lee, of Los Angeles, California. Thanks also to: Joy Renee Grieco, of Park Ridge, New Jersey.

What Do Federal Express Delivery People Do After 10:30 A.M.?

Federal Express is justly famous for its pledge to deliver Priority One packages before 10:30 A.M. the next business day. What, then, do delivery people do after the last priority package is delivered? Take a siesta? Smoke cigars? Play poker with U.S. Post Office employees who haven't delivered their first-class mail yet?

Actually, Federal Express keeps its employees hopping all day long. In some cities, packages are delivered as early as 7:30 A.M. (The pickup and delivery cycles of packages tend to be earlier in the West, because *all* packages are routed through Federal Express headquarters in Memphis, Tennessee prior to shipment to their eventual destinations.) Before any packages can be delivered, they must be sorted by routes; in smaller stations, the courier often does the sorting himself.

After all Priority One packages are delivered, the courier tries to drop off all second-day deliveries before noon. If he succeeds, he is likely to take a lunch break around midday.

After lunch, the pickup cycle begins. By the time the courier has gathered all the incoming packages, he has worked a full day. If there is any spare time at all, paper work has a way of filling it.

In large stations, the process of sorting routes, delivering Priority One packages, delivering second-day packages, picking up all packages, and filling out paper work can consume more than eight hours. For this reason, about 25 percent of all Federal Express employees are part-timers, often used for sorting packages for delivery by couriers. When a Federal Express courier drops off his last Priority One package before the 10:30 A.M. deadline, his workday has just begun.

Submitted by Merle Pollis, of Cleveland, Ohio.

Why Do So Many Mass Mailers Use Return Envelopes With Windows?

It's easy to figure out why many mass mailers (such as utilities or credit card companies) use window envelopes for the bills they send to you. Bills are prepared by computers and are stuffed into envelopes by inserting machines. The window eliminates the costly process of addressing each envelope separately.

But why the window on the return envelope? Couldn't the companies simply preprint their address, avoiding the problem of customers inserting the reply portion of the statement upside down or wrong side out?

There is a good reason for window reply envelopes. Many large companies use various geographical locations for receiving remittances. The window envelope saves the company the cost of printing several different addresses on reply envelopes.

Although the dire warnings on the back or flap of envelopes have almost eliminated the problem of incorrectly stuffed reply stubs, Pavey Envelope and Tag Corporation, of Jersey City, New Jersey, has recently developed the idea of clipping a corner on the return stub and gluing a corner of the envelope so that the stub can be inserted only one way—the correct way.

Submitted by Pat O'Conner, of Brooklyn, New York.

Why Does the Skin on the Extremities Wrinkle After a Bath? And Why Only the Extremities?

Despite its appearance, your skin isn't shriveling after your bath. Actually, it is expanding.

The skin on the fingers, palms, toes, and soles wrinkles only after it is saturated with water (a prolonged stay underwater in the swimming pool will create the same effect). The stratum corneum—the thick, dead, horny layer of the skin that protects us from the environment and that makes the skin on our hands and feet tougher and thicker than that on our stomachs or faces—expands when it soaks up water. This expansion causes the wrinkling effect.

So why doesn't the skin on other parts of the body also wrinkle when saturated? Actually, it does, but there is more room for the moisture to be absorbed in these less densely packed areas before it will show. One doctor we contacted said that soldiers whose feet are submerged in soggy boots for a long period will exhibit wrinkling all over the covered area.

Submitted by Michelle L. Zielinski, of Arnold, Missouri.

Submitted by Marley Sims, of Van Nuys, California. ▶

What Happens to the Razor Blades That Are Thrown Down Used-Blade Slots in Hotels?

Absolutely nothing. They are left to collect indefinitely between the studs of the walls. If you ever try to put a used blade down the razor slot and find the slot stuffed to the gills, you may assume that you are either in a very old hotel or in one that caters to a particularly hirsute clientele. Of course, there are fewer of those blades being deposited than there were years ago, because disposable razors are particularly popular among travelers; disposables won't fit in the skinny opening.

    If our civilization goes the way of the dodo and the remnants of our culture are buried in layers, we will certainly have some nasty surprises for future archaeologists. Between used razor blades and pop top can tabs, we will literally keep future diggers on their toes, or at least in very durable shoes.

Why Doesn't Evaporated Milk Have to Be Refrigerated?

Evaporated milk, of course, is thickened solely by evaporation; it is often confused with condensed milk, which is made by evaporating some of the cow's milk and adding sugar. Evaporated milk has a long shelf life because it is sterilized in the can, a steam-heat process that destroys potentially harmful microorganisms. Evaporated milk often develops a darkish off-color after about a year, but it is still safe to consume.

Submitted by Cassandra A. Sherrill, of Granite Hills, North Carolina.

Why Is Evaporated Milk Sold in Soldered Cans?

Can openers, and the people who use them, have difficulty with soldered cans. Is there a real advantage to using them?

Soldered cans are stronger than regular aluminum or tin cans. As we have just learned, evaporated milk is actually sterilized in the can, and manufacturers have found soldered cans to be more dependable and durable during the intense heating process. Marsha McLain, of Pet, Inc., told us that all of their cans are welded with double seams in two pieces. The bottom and sides are actually one piece, and are filled with liquid. Only after the milk is put into the container is the top soldered on.

What Causes the Ringing Sound You Get in Your Ears?

Unless you are listening to a bell, a ringing sensation means you are suffering from tinnitus. Someone with tinnitus receives auditory sensations without any external auditory source. While most of us rarely experience tinnitus, it is a chronic problem for over 30 million Americans.

Tinnitus is a symptom, not a disease in itself. Virtually anything that might disturb the auditory nerve is capable of causing tinnitus. Because the function of the auditory nerve is to carry sound, when the nerve is irritated for any reason the brain interprets the impulse as noise.

Some of the most common causes for temporary tinnitus are:

1. Reaction to a loud noise.
2. Vascular distress after a physical or mental trauma.
3. Allergic reaction to medication. (Aspirin is the most common pharmaceutical cause of tinnitus. Many people who take more than twenty aspirin per day are subject to tinnitus attacks). Luckily, the symptoms usually disappear upon discontinuance of the drug.

Causes of more chronic tinnitus conditions are myriad. Here are some of the most common: clogging of the external ear with earwax; inflammation of any part of the ear; drug overdoses; excessive use of the telephone; vertigo attacks; nutritional deficiencies (particularly a lack of trace minerals); muscle spasms in the ear; infections; allergies.

Chronic tinnitus sufferers have to live not only with annoying buzzing, but usually with accompanying hearing loss. Unfortunately, there is no simple cure for the condition. Much research is being conducted on the role of nutrition in helping treat tinnitus, but for now, the emphasis is on teaching sufferers how to live with the problem. Devices are sold to mask the ringing sound. Techniques such as hypnosis and biofeedback are used to distract the patient from the annoying ringing.

Ear problems may not be the most glamorous medical problems, but they are the most prevalent, as a booklet from the House Ear Institute, prepared by the Otological Medical Group, Inc., of Los Angeles, explains: "Loss of hearing is America's largest, yet least recognized, physical ailment. More people suffer from it than heart disease, cancer, blindness, tuberculosis, multiple sclerosis, venereal disease, and kidney disease combined."

Submitted by Bobby Dalton, of Maryland Heights, Missouri.

HOW Did Chocolate Bunnies for Easter Come About?

No doubt, the chocolate bunny was introduced for the same reason that candy corn was introduced for Halloween—in order to make more money for the candy industry. Purveyors of nonessential gift items (flowers, greeting cards, candy) are always looking for new reasons to compel customers to buy their products. If one were inclined toward conspiracy theories, one could look on everything from Mother's Day to National Secretary's Week as nothing but blatant attempts to pry discretionary dollars from hapless citizens.

Chocolate bunnies date back to the 1850s in Germany. Along with bunnies, chocolatiers sold chocolate eggs and chickens. Switzerland, France, and other European chocolate producers followed soon after. Most of the chocolate companies we contacted felt that the bunnies symbolized renewal and rejuvenation, and were intended to symbolize the "Rites of Spring," not strictly Easter. As Charlotte H. Connelly of Whitman's Chocolates told us, the chocolate bunnies spread rapidly to the United States from Europe.

At present, chocolate eggs and bunnies help bridge the "chocolate gap" that befalls the confectionary industry between St. Valentine's Day and Mother's Day.

Why Do Old Women Dye Their Hair Blue?

In the 1960s, it was fashionable to tint or bleach hair in pastel shades. Some older women, perhaps, are choosing to stick with a trend that has come and gone.

The majority, however, use a blue rinse (not a dye or tint) to combat the yellow shadings that discolor their gray or white hair. Blue helps mask yellow.

Advancing age is not the sole reason for yellow hair. Some chemicals used in other hair preparations can cause yellowing. But the biggest culprit of all is smoke. Cosmetician Richard Levac told *Imponderables* that as we get older, the hair becomes more porous. Smoke coats the hair and embeds itself in the hair shaft, causing yellowing.

Levac adds that very few women are intentionally trying to emerge from a salon with blue hair. Blue rinses are much lighter than they were twenty years ago. If you can notice the blue, the hairdresser has done a poor job.

Ironically, blue hair has now made a comeback of sorts with young girls, thanks to new wave music. And with Cyndi Lauper around, any primary color is fair game.

Submitted by Daniel A. Placko, Jr., of Chicago, Illinois.

What Are the Criteria for the Placement of a "Dangerous Curve" or "Dangerous Turn" Sign?

The answer comes from the encylopedic *Manual on Uniform Traffic Devices.* Individual states are free to deviate from the standards cited in this federal publication, but few do.

How do they determine if a turn is dangerous enough to warrant a warning sign? The *Manual*'s criterion for a turn sign (the black arrow at a right angle against a yellow background) is explicit:

> The Turn sign is intended for use where engineering investigations of roadway, geometric, and operating conditions show the recommended speed on a turn to be 30 MPH or less, and this recommended speed is equal to or less than the speed limit established by law or by that regulation for that section of a highway. Where a Turn sign is warranted, a Large Arrow sign may be used

on the outside of the turn. Additional protection may be provided by use of the Advisory Speed plate.

Note that these guidelines reflect the reality that actual traffic speeds usually exceed the law. Warning signs can be posted even if the "safe" speed is identical to the posted speed limit.

The criterion for the curve sign is similar: a curve sign can be placed any time tests demonstrate that the recommended speed should be between thirty and sixty miles per hour and that speed is equal to or less than the posted speed limit.

Submitted by Robert J. Abrams, of Boston, Massachusetts.

Why Don't We Ever See Dead Birds?

We see hundreds of birds on an average day, and occasionally spot one run over by a car, but why don't we ever see one dead from natural causes? Don't they ever keel over in flight? Do birds go someplace special to die?

Surprisingly, birds don't fly anywhere particular to die. The reason we don't see dead birds is that they are quickly scavenged by other animals. Although this sounds like a cruel fate, bird expert Starr Saphir views it differently, marveling at the efficiency of the natural world. The moment a bird can no longer function, it is used as valuable fuel. Birds are eaten by cats, dogs, rats, opossums, small insects, and even bacteria. Saphir told us that she has led birdwatching walks and seen the intact but dead body of a bird on the ground on the first leg of the walk; on the way back, an hour later, the majority of the body was already scavenged. Within twenty-four hours, the remains of most birds, in the wild or in an urban area, would presumably become only a pile of feathers.

Richard C. Banks, vice-president of the U.S. Ornithologist's Union, told *Imponderables* that a few birds might actually die in flight (although he had not personally ever seen this happen). The most likely candidates would be migrating birds flying over the ocean, far away from food sources and without convenient landing spots to fight off exhaustion. Sick birds generally don't take wing in the first place.

Submitted by Cecilia F. Boucher, of Roslindale, Massachusetts. Thanks also to: Walter Bartner, of New York, New York; Thomas Cunningham, of Pittsburgh, Pennsylvania; L. T. Quirk, of Red Bank, New Jersey; and Richard Rosberger, of Washington, D.C.

Why Do All Packaged Bakery Goods Seem to Be Registered by the Pennsylvania Department of Agriculture?

In 1933, the state of Pennsylvania passed a Bakery Inspection Act mandating that all bakery goods must be inspected and registered in order to be sold in the state. Further, no packaged bakery goods could be sold in Pennsylvania unless the registration notice was printed on its wrapping. The law was enacted to ensure not only the wholesomeness of the food but the accuracy of the weight stated on the package and the health of the employees handling the food (all bakery employees must have an annual physical examination).

It is easy to understand why Pennsylvania would want to protect the welfare of its citizens, but why are Pepperidge Farm cookies, Hostess cupcakes, Wonder bread, and other nationally distributed baked goods also registered by the Pennsylvania Department of Agriculture? Because the law does not exempt out-of-state bakeries from having to print "Registered by the Pennsylvania Department of Agriculture" on its packaging —without it, the goods cannot be sold anywhere in Pennsylvania. Instead of going to the extra expense of printing separate wrappers for the state of Pennsylvania, manufacturers include its registration on their labels all across the country.

How does Pennsylvania monitor the wholesomeness in bakeries out of state or out of country? According to Dick Elgin, of the Pennsylvania Department of Agriculture, the state has reciprocal agreements with food inspection units in other states. Most states have laws regulating the wholesomeness of bakeries; it is only the requirement to print the state's "seal of approval" that differentiates Pennsylvania.

Meat, poultry, and eggs are the only foodstuffs that require inspections by the federal government. While the Pennsylvania Department of Agriculture registration won't promise you great taste or even good nutrition, it will reassure you that the plant where your cookie was baked was inspected at least once a year and that some inspector lived after popping a similar cookie into his or her mouth.

Submitted by Carol Jewett, of New York, New York.

How Do They Keep All the Raisins in Cereal Boxes from Falling to the Bottom?

The Rule of Popcorn Physics, which states that unpopped popcorn kernels fall to the bottom of the bowl, has saved many a tooth for generations. The explanation for this immutable law is easy enough to comprehend: unpopped kernels fall to the bottom both because their density is greater than expanded popcorn and because our handling of the corn creates crevices for the unpopped kernel to slide down.

Many inquisitive types have searched for corollaries to the Popcorn Physics rule. For example, the tenet of Slithery Sundaes posits that regardless of how much syrup or toppings one puts atop ice cream in a sundae, it will all fall to the bottom of the bowl anyway, collecting in a pool of glop.

So it was not without a feeling of reverence and awe that we approached the subject of raisins in cereal boxes, tiny dried grapes that seem to defy the usual laws of food gravity. Linda E. Belisle, at General Mills, supplied the simple but elegant solution.

Raisins are added to boxes only after more than half of the cereal has already been packed. The cereal thus has a chance to settle and condense. During average shipping conditions, boxes get jostled a bit (the equivalent of our stirring the contents of a popcorn bowl while grabbing a fistful), so the raisins actually sift and become evenly distributed throughout the box.

The tendency of cereal to condense within the package is responsible for the warning on most cereal packages that the contents are measured by weight rather than volume. Little did you know that this condensation was also responsible for the Law of Rising Raisins.

Submitted by James A. Hoagland, of Stockton, California.

Why Do Runs in Stockings Usually Run Up?

A complicated issue, it turns out, but one that the folks at Hanes and L'eggs were happy to tackle. The direction in which runs will go is determined by the type of stitching used in the construction of the hosiery. The leg portions of most panty hose and sheer nylons are woven in what is called the "jersey stitch" or "stocking stitch." The jersey stitch is produced by one set of needles when all of the needles produce plain stitches at every course. Hosiery made from jersey stitches runs or "ladders" both up and down.

Most manufacturers use the jersey stitch for their basic panty hose and stocking styles. Jersey stitches provide a smoother feel and a sheerer look than other constructions, yet they are still durable and stretch well.

Other often-used stitches include the "run resist," the "float," and most popular, the "tuck," all of which *will only run up*. L'eggs, for example, uses the tuck stitch on their control-top panties. When the yarn in the stitch is severed, it will only run upward. The purpose, according to L'eggs, is "to prevent the run from encroaching onto the part of the hose that you can see."

Why don't the manufacturers always use a stitch that will ladder up, then, as this construction will most often prevent the run from being visible? Hanes Hosiery's answer is that tucks, run-resist, and float stitches all feel rough on the leg and look heavier on the leg than the jersey stitch. Most manufacturers use the float and tuck stitches for stockings that are designed to look heavier, particularly patterned and mesh hosiery.

Submitted by Sara Vander Fliet, of Cedar Grove, New Jersey.

Does Putting Women's Hosiery in the Freezer Forestall Runs?

On one thing L'eggs and Hanes can agree. Despite all folk wisdom and advice columns to the contrary, putting hosiery in the freezer does not forestall runs. Mary S. Gilbert of L'eggs states that "hosiery is made of synthetic fibers which are not affected by cold."

Eleanor Pardue, product evaluation manager at Hanes, was familiar with the nylon in the freezer claim, but remains firm:

> Based on the physical testing I am familiar with, there is no difference in the breaking strength of nylon which has never been frozen and nylon that has been frozen. I do know that one can wear two identical pairs of stockings manufactured at the same time under identical circumstances and one pair may run the first time worn while the other pair may last through ten wears.

Submitted by Bonnie Gellas, of New York, New York.

Why Do Traffic Signals Use Red, Yellow, and Green Lights? Why Is the Red Light on Top, Green Light on the Bottom, and Yellow Light in Between?

Traffic signals actually predate the existence of the automobile. One was installed outside of British Parliament in 1868. This signal (and some early American variations) had two semaphore arms, like a railroad signal, that acted as a physical impediment to oncoming traffic.

The English device was designed to control the flow of pedestrians, and some feature was needed to make it functional at night. The easiest solution was to adapt the system used for railroad signals—red and green gas lamps would signify when one could proceed (green) or had to stop (red). This British prototype wasn't a rousing success—it blew up shortly after its introduction, killing a London policeman.

A lively controversy has developed over where the first modern traffic signal designed to control automobile traffic was in use. Although Salt Lake City and St. Paul lay claim to the crown, the green-red signal installed on Euclid Avenue in Cleveland, Ohio, in 1914 is generally credited with being the first.

Although the traffic signal's colors might have been arbitrarily lifted from the railroad's, there is an important safety reason for the consistency of the configuration today. As recently as the 1950s, many traffic signals, especially in busy urban intersections, were displayed horizontally rather than vertically. The current vertical design with red on top was adopted in order to aid color-blind individuals who might be confused by different layouts. According to Eugene W. Robbins, president of the Texas Good Roads/Transportation Association, the red in traffic signals has some orange in it and the green has some blue in order to make it even easier for the color blind to distinguish them.

Submitted by John Branden, of Davis, California. Thanks also to: Maya Vinarsky, of Los Angeles, California; Sean Gayle, of Slidell, Louisiana; Eddie Haggerty, of Waseca, Minnesota; William Debovitz, of Bernardsville, New Jersey.

THIS SIDE FOR MILKING

SEE UDDER SIDE

Why Are Cows Usually Milked from the Right Side?

Although this subject is usually not part of the veterinary school curriculum, we went right to the organization best suited to answer the Imponderable: the American Association of Bovine Practitioners and its officer Dr. Harold E. Amstutz. Although Dr. Amstutz said he had never considered this question before we posed it, he was ready with a sensible explanation:

> Since most people are right handed, it is more logical to sit down on the right side of the cow and have more room to maneuver the milk bucket with the right hand between the cow's front and rear legs. There would not be nearly as much room to maneuver the bucket with the right hand if a right handed person were to sit on the left side.
>
> In general, we think of "right" as correct and "left" as being wrong. Cows have no preference since we milk them from either

side in today's milking parlors. The only ones that would have a preference are those that were trained to be milked on one side and then someone tried to milk them from the other side. The milker would probably be kicked in that case.

Submitted by Marci Perlmutter, of Warren, New Jersey.

Why Do Many Merchants Ask Customers to Put Their Addresses and Phone Numbers on Credit Card Slips?

If Visa or MasterCard or American Express needed your address or phone number, wouldn't they reserve a spot on their credit-card slips for them? Clearly, it is the stores that want such data.

But why do the stores need this information? If you charge goods on a stolen credit card, it is the creditor, not the merchant, who gets stuck with the bill, as long as the merchant complies with all of the security arrangements (such as verifying all purchases above a certain amount).

The credit-card companies themselves couldn't think of a good reason for stores to ask for address and phone number. After all, if somebody is going to steal a credit card, he is unlikely to provide accurate instructions on how to locate him. Perhaps, one credit-card executive speculated, the stores make this request in order to compile a mailing list.

So we talked to merchants. Their reasons turned out to be prosaic. Some of them mentioned that they carry insurance on bad checks, and that part of the agreement with insurance companies is that all customers must supply addresses, phone numbers, and driver's license or social security number. If they don't include these data, merchants aren't reimbursed for bad checks. Although they realize that such precautions are irrelevant when it comes to collecting money from credit cards, some merchants believe that by forcing employees to collect addresses and phone numbers from all noncash customers, clerks would be less likely to forget to ask for the information from people using checks.

All of the merchants in New York City gave one reason, and several mentioned it as the only reason, for asking customers to include address and phone numbers, and this explains why the practice is uncommon in small towns. With the information provided they can contact customers when they leave their credit card at the store by mistake!

Submitted by Mark Schulman of Altamonte Springs, Florida.

On Airplanes, Why Do Our Ears Pop and Bother Us More on Descent Than on Ascent?

The ear is composed of three parts:

1. The outer ear, which includes the part of the ear that is visible, plus the ear canal connected to the eardrum.
2. The middle ear, which includes the eardrum, the ear bones (ossicles), and the air spaces behind the eardrum and in the mastoid cavities.

3. The inner ear, which contains the nerve endings that facilitate hearing and equilibrium.

The middle ear is what bothers travelers on airplanes because it is, in part, an air pocket vulnerable to changes in air pressure. On the ground, when you swallow, your ears make a little click or popping sound. This noise marks the passage of a small air bubble up from the back of your nose, through the eustachian tube, and into your middle ear. According to the American Council of Otolaryngology, "the air in the middle ear is constantly being absorbed by its membranous lining, but it is frequently re-supplied through the eustachian tube during the process of swallowing. In this manner air pressure on both sides of the eardrum stays about equal. If, and when, the air pressure is *not* equal, the ear feels blocked."

If the eustachian tube is blocked, no air can be replenished in the middle ear; any air present absorbs and a vacuum occurs, sucking the eardrum inward. Blocked eustachian tubes can cause a loss of hearing and pain.

A clear and properly functioning eustachian tube is the key to problem-free ears on plane flights; if it can open wide enough and often enough, the eustachian tube can moderate changing air-pressure conditions. When you ascend on an airplane, it is to less pressure, so the air expands in the middle ear. The eustachian tube works much like a flutter valve on an automobile. When you ascend, the air in your ear is forced through the tube in a steady stream without any problem.

When you descend, it is to greater air pressure. A vacuum forms even faster in the middle ear, making it harder for the air to go back through the membranous part of the eustachian tube. According to Dr. Andrew F. Horne, in the Office of Aviation Medicine of the Federal Aviation Administration, the ear popping is caused when the valve of the eustachian tube opens and closes. On ascent, the air runs through the eustachian tube in a steady stream; on descent, the air must contend with the membranous part of the eustachian tube. Without the steady air flow, it takes longer to equalize air pressure inside and outside your ear.

Airplane pilots are taught how to counteract differences in air pressure. The simple act of swallowing pulls open the eustachian tube, which is why gum chewing or candy sucking has become a takeoff and landing ritual for many passengers. Yawning is even more effective, for it pulls the muscle that opens the eustachian tube even harder than swallowing.

If neither swallowing nor yawning works, the American Council of Otolaryngology recommends this procedure:

1. Pinch your nostrils shut.
2. Take in a mouthful of air.
3. Using your cheek and throat muscles, force the air into the back of your nose as if you were trying to blow your thumb and fingers away from your nostrils.
4. When you hear a loud pop in your ears, you have succeeded, but you may have to repeat the process again during descent.

WHY DO CLOCKS RUN CLOCKWISE?

Where Do Houseflies Go During the Winter?

To heaven, usually. Some flies survive winter, but only under extremely favorable conditions, when they can take shelter in barns or inside human residences where they can find enough organic matter and warmth to eat and breed.

Even under the best of circumstances, the normal life-span of a housefly north of the equator is approximately seven to twenty-one days. The most important variable in the longevity of these insects is the ambient temperature—they die off in droves when it falls below freezing or becomes excessively hot.

Although they actually live longest in cool temperatures, because they are less active, flies breed most prolifically when temperatures are warm, food is abundant, and humidity is moderate. Winter tends to deprive them of all of these favorable conditions, so that they not only die off themselves, but do so without having been able to breed successfully. The U.S. Department of Agriculture claims that no housefly has been proved to live from autumn to spring (which answers another Imponderable: why do we see so few houseflies in the spring?).

So how can they regenerate the species? Most people believe that flies hibernate or become dormant, like some other insects, but this theory has proved to be untrue. The few flies that we find in the spring are mainly the descendants of the adult flies that managed to find good hiding places during the previous winter. These spring flies breed their little wings off, just in time to harrass you on your picnics when the weather gets good.

Some of the flies that survive winter are not adults, but rather flies in their earlier developmental stages. Fly eggs are usually deposited in the ground, in crevices, in wood, or in a particular favorite, cow manure. These eggs hatch, literally, in a few hours, and turn into larvae, a phase that can last anywhere from one to four days. Larvae feed on decaying plant or animal matter (such as other insect larvae). As the fly larva grows, it undergoes pupation, a phase that lasts about five days, in which the fly rests as its larval features are transformed into adult ones. Many entomologists used to answer this Imponderable by speculating that most flies that survive the winter do so in the form of larvae or pupae, but scientists now believe that adult flies have a much better chance of surviving the winter than their younger brethren, who have a hard time coping with cold weather. Still, some larvae and pupae do stay alive during the very end of winter and develop into adults in the spring.

The fecundity of the *Musca domestica* is truly awesome. One scientist estimated that a single mating pair of houseflies could generate as many as 325,923,200,000,000 offspring in one summer. One-sixth of a cubic foot of soil taken in India revealed 4,024 *surviving* flies. Maybe the Imponderable should read: why isn't the entire world overcome with flies?

Any notion that flies migrate south during the winter is easily dispelled. The average flight range of a housefly is a measly one-quarter of a mile. Scientists have tracked the flight of flies: they rarely go beyond a ten-mile radius of their birthplace during their entire lifetime.

Where Does White Pepper Come From?

From black pepper. The most popular of all spices (salt is not a spice) is not related to sweet red, green, or hot peppers, but is the dried berry of a woody, climbing vine known as *Piper nigrum L.*

On the vine the peppercorn is neither white nor black. As the fruit ripens, it turns from green to yellow and then to red. To make black pepper, the berries are picked while somewhat immature and then dried. As they dry, their skin turns a dark color. When ground, the pepper contains both light and dark particles —because the whole peppercorn is used—but the general appearance is dark.

White pepper is left on the vine to mature, at which point it is easier to separate the dark skin. The berries are soaked to loosen the skin as much as possible and then rubbed to remove it entirely. After the dark skin is discarded, the naked white peppercorns are put out in the sun to dry.

Technology has caught up with the spice world. Some white pepper, usually known as "decorticated white pepper," is now produced by removing the skin of dried *black* peppercorns by machine. Decorticated pepper looks like white pepper but tastes more like black pepper.

Why bother with white pepper? Often it is used solely for aesthetic purposes, such as in light-colored sauces and soups where little black specks may upset the chef's carefully orchestrated balance (or be misconstrued as little black insect fragments). Some spice wimps also prefer white pepper for its milder taste and smell.

Ted Turner does not have a monopoly on colorization. Go into any gourmet store and you will encounter green peppercorns. These immature berries are not left out in the sun but either packed in liquid (usually wine vinegar or brine) or freeze-dried in order to retain the distinctive green color. Because green peppercorns are harvested at an early stage of the berry's development, they are quite mild, but they do have a distinctive taste, which is prized by nouvelle cuisine restaurateurs.

Submitted by Kathy Cripe, of South Bend, Indiana.

What Purpose Do Wisdom Teeth Serve?

They serve a powerful purpose for dentists, who are paid to extract them. Otherwise, wisdom teeth are commonly regarded as being useless to modern man. But because nature rarely provides us with useless body parts, a little investigation yields a more satisfying answer.

Primitive man ate meats so tough that they make beef jerky feel like mashed potatoes in comparison. The extra molars in the back of the mouth, now known as wisdom teeth, undoubtedly aided in our ancestors' mastication.

As humans have evolved, their brains have gotten progressively larger and the face position has moved farther downward and inward. About the time that primitive man started walking in an upright position, other changes in the facial structure occurred. The protruding jawbones of early man gradually moved backward, making the jaw itself shorter and leaving no room for the wisdom teeth (also known as third molars). Most people's jaws no longer have the capacity to accommodate these four, now superfluous, teeth.

Why Are Ancient Cities Buried in Layers? And Where Did the Dirt Come From?

This Imponderable assumes two facts that aren't always true. First, not all ruins are the remains of cities. Many other ancient sites—such as forts, camping sites, cave dwellings, cemeteries, and quarries—are also frequently buried. Second, not all ancient cities are buried; once in a while, archaeologists are given a break and find relics close to or at the surface of the ground.

Still, the questions are fascinating, and we went to two experts for the answers: George Rapp, Jr., dean and professor of geology and archaeology of the University of Minnesota, Duluth, and coeditor of *Archaeological Geology*; and Boston University's Al B. Wesolowsky, managing editor of the *Journal of Field Archaeology*. Both stressed that most buried ruins were caused by a combination of factors. Here are some of the most common:

WHY DO CLOCKS RUN CLOCKWISE?

1. Wind-borne dust (known to archaeologists as "Aeolian dust") accumulates and eventually buries artifacts. Aeolian dust can vary from wind-blown volcanic dust to ordinary dirt and house dust.

2. Water-borne sediment accumulates and eventually buries artifacts. Rain carrying sediment from a high point to a lower spot is often the culprit, but sand or clay formed by flowing waters, such as riverine deposits gathered during floods, can literally bury a riverside community. Often, water collects and carries what are technically Aeolian deposits to a lower part of a site.

3. Catastrophic natural events can cause burials in one fell swoop, though this is exceedingly rare, and as Dr. Rapp adds, "In these circumstances the site must be in a topographic situation where erosion is absent or at least considerably slower than deposition." Even when a city is buried after one catastrophe, the burial can be caused by more than one factor. Dr. Wesolowsky notes that although both Pompeii and Herculaneum were buried by the eruption of Mt. Vesuvius in A.D. 79, one was buried by mudflow and the other by ashflow.

4. Manmade structures can collapse, contributing to the burial. Sometimes this destruction is accidental (such as floods, earthquakes, fires), and sometimes intentional (bombings, demolitions). Humans seem incapable of leaving behind no trace of their activities. Says Rapp: "Even cities as young as New York City have accumulated a considerable depth of such debris. Early New York is now buried many feet below the current surface."

5. Occasionally, ancient civilizations did their own burying. Wesolowsky's example:

> When Constantine wanted to build Old St. Peter's on the side of the Vatican Hill in the early fourth century, his engineers had to cut off part of the slope and dump it into a Roman cemetery (thereby preserving the lower part of the cemetery, including what has been identified as the tomb of Peter himself) to provide a platform for the basilica. When Old St. Peter's was demolished in

the sixteenth century to make way for the current church, parts of the old church were used as fill in low areas in the locale.

Rapp's example:

> This phenomenon is best seen in the tels of the Near East. Often they are tens of feet high. Each "civilization" is built over the debris of the preceding one. The houses were mostly of mud brick, which had a lifetime of perhaps sixty years. When they collapsed the earth was just spread around. In two thousand or three thousand years these great habitation mounds (tels) grew to great heights and now rise above the surrounding plains. Each layer encloses archaeological remains of the period of occupation.

While we self-consciously bury time capsules to give future generations an inkling of what our generation is like, the gesture is unnecessary. With an assist from Mother Nature, we are unwittingly burying revealing artifacts—everything from candy wrappers to beer cans—every day.

Submitted by Greg Cox, of San Rafael, California.

What's the Difference Between an X-Rated Movie and an XXX-Rated Movie? Why Isn't There an "XX" Rating?

The Motion Picture Association of America issues the movie ratings you see in the newspaper. Motion picture companies are under no legal obligation to have their movies rated, but they are not allowed to affix their own rating. In order to obtain a G, PG, PG-13 or R rating, a fee must be paid to the MPAA. An MPAA committee views each film and issues an edict that sets the rating, subject to appeal. None of the major film companies is willing to bypass the MPAA ratings. Since the rating codes were instituted in the 1960s, there has actually been much less pressure on the studios to reduce violence and sexual content. Also, some newspapers refuse to accept advertising for non-MPAA-rated movies, and most film executives feel that the rating system has worked reasonably well as a warning device for concerned parents.

The X-rating was originally conceived as the designation for any movie suitable only for adults, regardless of genre. Such critics' favorites as the Best Picture Oscar-winning *Midnight Cowboy* were rated X because of their mature subject matter, and *A Clockwork Orange* was rated X for its violence and intensity.

With only a few other exceptions, nonpornographic X-rated movies have bombed at the box office. Any film that catered to adults automatically excluded many of the most rabid moviegoers—teenagers. The advertisements for so-called "adult films" gladly trumpeted their X ratings: how better to prove the salaciousness of a movie than by prohibiting children from viewing it? Even better, MPAA rules allowed companies to rate their films X without the association's certification, a policy that enabled low-budget film companies to nab an X rating without paying the fee of nearly a thousand dollars. As the few mainstream X-rated films were overwhelmed by the multitude of X-rated porn movies, major film companies like Paramount and Columbia refused to release any X-rated movies, for X had become synonymous with smut.

The producers of adult films had the opposite problem. Here they were, trying to purvey their X-rated product, when prestigious films like *Midnight Cowboy* were sullying the reputation of the adults-only rating by containing redeeming social value.

David F. Friedman, board chairman of the Adult Film Association of America, told us that the XXX rating was actually started as a joke, to distinguish "straight films," with mature content, from pornography. There is not now and has never been a formal XXX rating for movies; it has always been a marketing ploy adopted by film distributors and/or movie exhibitors.

Is there any difference between an X- and an XXX-rated movie? According to Friedman, no. Although some customers might believe that an XXX-rated movie is "harder" than the simple X, this has never been the case. Many pornographic films are made in several versions: hard-core X-rated; a "soft" X, used for localities where hard-core is banned; a "cable" version, a doctored once-explicit version; and an expurgated R-rated version, designed for playoffs in nonporno theaters, such as drive-ins. Whether or not any of these versions of a pornographic movie is billed as X or XXX is more dependent on the whims of the producer or the theater management than on the content of the movie.

Why no XX rating? Who knows? Once someone started the XXX, who was going to say that their movie wasn't quite as sexy? X-inflation is likely to remain rampant as long as there are pornographic theaters.

Submitted by Richard Rosberger, of Washington, D.C. Thanks also to: Curtis Kelly, of Chicago, Illinois, and Thomas Cunningham, of Pittsburgh, Pennsylvania.

Where Does a New Speed Limit Begin? Does It Start at the Speed Limit Sign, at Some Point Beyond the Sign, or Where the Sign Becomes Clearly Visible?

If a speed limit drops from fifty-five miles per hour to thirty-five miles per hour, isn't it clearly legal to drive at fifty-five miles per hour until you pass the thirty-five miles per hour sign? But how are we expected to drop twenty miles per hour instantaneously? Is there a grace period, a distinct length of road on which we are exempt from the new speed limit?

No such luck. The speed-limit sign is posted precisely where the new limit takes effect. How you slow down to the new speed is your business, and your problem.

Of course, traffic laws are up to the individual states, but most legislatures rely on the provisions of the federal government's *Manual on Uniform Traffic Control Devices*. And the manual is unambiguous: "Speed limit signs, indicating speed limits for which posting is required by law, shall be located at the points of change from one speed limit to another. . . . At the end of the section to which a speed limit applies, a Speed Limit sign showing the next speed limit shall be erected." The one provision intended to help drivers slow down before a new speed limit is the "Reduced Speed Ahead" sign. These are placed primarily in rural areas where drops in speed limits can easily reach twenty to thirty-five miles per hour. But these warning signs must be followed by a speed-limit sign that marks precisely where the altered speed limit applies.

Submitted by Glenn Worthman, of Palo Alto, California.

If the National Speed Limit is 55 Miles per Hour, Why Do Speedometers Go up to 85 Miles per Hour and Higher?

The Department of Transportation mandated the maximum speedometer reading effective September 1, 1982. The rule read:

> No speedometer shall have graduations or numerical values for speeds greater than 140/km/h and 85 mph and shall not otherwise indicate such speeds. Each speedometer shall include "55" in the mph scale. Each speedometer, other than an electronic digital speedometer, shall highlight the number 55 or otherwise highlight the point at which the indicated vehicle speed equals 55 mph.

Benn Dunn, manager of product technical communications for American Motors, says that the National Highway Traffic Safety Administration offered two reasons for the 85 mile per hour limit. First, the limit would allow speedometer dials to be more precisely graduated and more readable in the range of reasonable driving ranges. Second, the upper limit presumably "reduced the temptation for immature drivers to test the upper speeds of their vehicles on public roads."

The regulation simply didn't work. Although there are no current federal regulations concerning what speeds should be shown on speedometers, all of the big four automakers continue, voluntarily, to maintain 85 miles per hour as the maximum speed indication on the analogue speedometers of most of their cars. Mr. Dunn predicted, however, that we will soon begin to see higher markings on analogue speedometers.

On their high-performance cars, U.S. automakers all exceed the 85 mile per hour standard. Obviously, the auto companies are not trying to encourage reckless driving, but the speedometer with a 125 mile per hour capacity is an effective marketing ploy. A car with a high maximum reading sends a message to the consumer that the car must be capable of attaining these speeds.

The automakers give three main reasons why it is important to maintain indications beyond the federal speed limit:

1. As P. M. Preuss, of Ford Motor Company, explained it, "Car speedometers are labeled beyond 55 mile per hour speeds because people drive in excess of 55 miles per hour. Obviously, some of these drivers are reckless, but by no means all. Automakers are under no obligation to produce cars that can go only 55 miles per hour; drivers who exceed the legal limit should be aware of how fast they are going." Law-abiding citizens exceed the speed limit under many circumstances. Passing maneuvers often require bursts of speed for brief periods of time. Drivers approaching a steep upgrade, reasonably enough, want to gather a head of steam before the climb. And drivers entering expressways must often speed up for their own and others' protection.

2. Speed limits can change. Particularly in rural areas, the 55 mile per hour limit has never been accepted and has been viewed as an affront to basic liberties.

3. Automobile engineers need speedometers with more generous indications. Many of the procedures they use to assess the safety and performance of cars, including tire, brake, and component tests, are carried out at speeds greater than 55 miles per hour.

Soon, analogue speedometers will probably give way to electronic speedometers. At present, electronic models are in short supply, so they are primarily a luxury option. Electronic speedometers feature continuous digital readouts, usually in two mile per hour increments, that register accurately whatever speed the car is traveling, regardless of the speed limit.

Submitted by Daniel C. Papcke, of Lakewood, Ohio.

What Is the Purpose of Pubic and Underarm Hair, the Only Body Hair That Men and Women Share in Abundance?

Even though humans have lost most of their fur, pubic hair and underarm hair remain in both sexes (at least, in most of the world —the majority of American women shave their armpits, for some reason we at *Imponderables* are still trying to ascertain). Any logical reason for this?

The most popular explanation is that pubic hair and armpit hair both trap the milky fluid secreted by the sebaceous glands. When the secretion is broken down by bacteria, a strong odor that acts as an aphrodisiac is generated. Isn't it ironic, then, that deodorants and antiperspirants are trumpeted for their ability to mask offensive odors? We are so worried about carrying bad smells that we neglect to realize that body odor can attract others. Perhaps deodorants should be marketed for people who *want* to get rid of the opposite sex.

Zoologists offer another explanation for pubic hair. Many animals, especially primates, have striking visual features around their genitals to help attract potential mates (have you seen a baboon lately?). The wide patch of pubic hair on an otherwise naked skin might have remained on humans for the very same reason.

Submitted by Barbara and Celeste Hoggan, of El Paso, Texas.

Why Do Construction Crews Put Pine Trees on Top of Buildings They Are Working On?

The tree atop buildings (and bridges) under construction is known as the "topping out" tree and celebrates the completion of the basic skeleton of the structure. In skyscrapers, an evergreen is attached to the top beam as it is hoisted, a signal that the building has reached its final height. For some builders, the evergreen symbolizes that none of the construction crew died in the effort. For others, the tree is a talisman for good luck and prosperity for the future occupants of the building.

While the topping-out ceremony of today is often accompanied by a celebration, complete with boring speeches by local politicians and the popping of flashbulbs, the precursors of topping out are ancient. Like many of our benign rituals, topping-out celebrations stem from ancient superstitions. The Romans marked the completion of the Pons Sublicius over the Tiber River in 621 B.C. by throwing some humans into the river as a sacrifice to the gods. While we now launch ships by banging them with champagne bottles, a different liquid—human blood—was used in earlier times. In ancient China, the ridgepoles of new buildings were smeared with chicken blood in an attempt to fool the gods into believing they were receiving the human counterpart. Many cultures feared that evil spirits occupied new structures, so well into the Middle Ages, priests and rabbis performed special blessings on new homes and public buildings.

The first evidence of trees being hoisted atop buildings was in A.D. 700 in Scandinavia, when they signaled that a completion party was about to begin. Black Forest Germans celebrated the nativity of Jesus Christ with the hoisting of Christmas trees. Today, topping-out trees are still most prevalent in northern Europe, particularly Germany and the Scandinavian countries. Indeed, Scandinavia's greatest playright, Henrik Ibsen, had his protagonist in *The Master Builder* meet his doom by falling while placing a topping-off wreath on one of his new buildings.

The evergreen has been joined by the stars and stripes as a topping-off symbol in the United States. According to *The Ironworker* magazine, "When the last strands of cable were laid for the Brooklyn Bridge a hundred years ago, the wheel operated by Ironworkers was decorated with American flags. By 1920 ironworkers were again draping their work with American flags, this time while driving the first rivet on the Bank of Italy in San Francisco." Flags atop buildings today signify not only the patriotism of the construction crew but also, in some cases, that the building was financed by public funds.

Submitted by Robert J. Abrams, of Boston, Massachusetts.

Why Aren't Whitewall Auto Tires as Wide as They Used to Be?

The sole purpose of white sidewall tires is to look pretty. At one time, whitewalls were all the rage and an option that most Americans bought. A caller on a radio talk show, who posed this Imponderable, questioned why, although he paid a hefty premium for the whitewalls, he got much less white for his money than he did years ago.

Talks with tire experts yielded two explanations. First, as whitewalls are a totally cosmetic option, their appearance is subject to the whims of fashion. Tire design, like high-fashion design, tends to follow the lead of Europeans. Porsche has evidently made a huge impact with its all black tires. Whitewalls are not particularly hip at the moment, so a flashy display of white on the sidewalls at this time is as likely to impress the opposite sex as a panoply of gold chains.

Second, the thickness of white sidewalls has been reduced to conform to a general decrease in the thickness of tires and tire components, part of an industrywide attempt to make tires run cooler in order to meet the requirements of high speeds on interstate highways and of federal high-speed safety standards.

The extra charge for whitewalls actually does reflect higher expenses in manufacturing them. According to Firestone Tire & Rubber Company, white rubber is slightly more expensive than black rubber per pound, but there are two other factors that increase the cost of whitewalls: extra time and steps are necessary to manufacture and finish whitewalls, and the black rubber adjacent to the white rubber must be treated to keep the white from being stained by what would be a normal migration of materials within the tire.

Why Do Clocks Run "Clockwise"?

In baseball, horse racing, and most forms of skating, we are accustomed to seeing a counterclockwise movement. Is there any particular reason why clocks run "clockwise"?

Henry Fried, one of the foremost horologists in the United States, gives a simple explanation for this Imponderable. Before the advent of clocks, we used sundials. In the northern hemisphere, the shadows rotated in the direction we now call "clockwise." The clock hands were built to mimic the natural movements of the sun. If clocks had been invented in the southern hemisphere, Fried speculates, "clockwise" would be the opposite direction.

Submitted by William Rogers, of St. Louis, Missouri.

On Clocks and Watches with Roman Numerals, Why Is Four Usually Noted as IIII Rather than IV?

Watch and clock designers are given great latitude in designating numbers on timepiece faces. Some use arabic numbers, most use roman numerals, and a few use no numbers at all.

But have you noticed that while the number nine is usually designated as "IX" on timepieces, four is almost universally designated as "IIII"? We contacted some of the biggest manufacturers of watches, and even they couldn't pinpoint the derivation of this custom. But they sent us to our friend, Henry Fried, who swatted away this Imponderable as if it were a gnat.

When mechanical clocks were first invented, in the fourteenth century, they were displayed in public places, usually on cathedrals. The faces themselves were only ornamental at first, for the early models had no hour or minute hand but merely gonged once for every hour of the day.

Clocks were thus of special value to the common people, who were almost universally illiterate. Most peasants, even in Italy, could not read roman numerals, and they could not subtract. They performed calculations and told time by counting on their fingers. Four slash marks were much easier for them to contend with than "IV," taking one away from five.

Many early clocks displayed twenty-four hours rather than twelve. While some German clocks in the fifteenth and sixteenth centuries used roman numerals to denote A.M. and arabic numbers for P.M., all-day clocks remained especially troublesome for the illiterate. So some clock designers always displayed all numbers ending with four or nine with slash marks rather than "IV" or "IX."

Why do clockmakers persist in using roman numerals today? Primarily because the touch of antiquity pleases consumers. At a time when dependable clocks and watches can be produced for less than they could decades ago, manufacturers need design elements to convince consumers to spend more. Although some argue that roman numerals are easier to read upside down and at a distance, the touch of class they connote is still their biggest selling point.

The delicious irony, of course, is that this touch of class stems from a system designed for peasants.

Why Are Rain Clouds Dark?

Rain is water. Water is light in color. Rain clouds are full of water. Therefore, rain clouds should be light. Impeccable logic, but wrong.

Obviously, there are always water particles in clouds. But when the particles of water are small, they reflect light and are perceived as white. When water particles become large enough to form raindrops, however, they absorb light and appear dark to us below.

Why Are So Many Corporations Incorporated in Delaware?

We blanched when we noticed that two of the largest New York banks, Citibank and Chase Manhattan, were incorporated in Delaware. Both banks' names betray their New York roots, so surely there must be some practical reasons why they chose to incorporate in another state.

Then we encountered a November 1986 *Forbes* article, which reported that Delaware houses more than thirty out-of-state banks. A call to the Delaware Chamber of Commerce yielded even more startling statistics. More than 170,000 companies are incorporated in Delaware, including more than one-half of all Fortune 500 companies, 42 percent of all New York Stock Exchange listees, and a similar proportion of AMEX companies.

How could Delaware, the home of fewer than 700,000 people, house so many corporations? The answer is a textbook illustration of the ways a small state can attract big business by changing its laws and tax structure to attract outsiders. One of the reasons that Delaware attracted so many banks, for example, is that it abolished usury ceilings, which are set by the state rather than by the federal government. Let's look at the other inducements that Delaware offers corporations seeking a home.

Favorable Tax Laws

1. No state sales tax.
2. No personal property tax.
3. No corporate income tax for corporations maintaining a corporate office in Delaware but not doing business in the state. If Chase Manhattan were incorporated in New York, New York State would demand a share of the income generated beyond its borders.
4. No corporate income tax for holding companies handling intangible investments or handling tangible properties located outside Delaware.
5. An extremely low franchise tax, based on authorized capital stock (the minimum is a staggeringly low $30; but there is also a maximum, $130,000 per year, that is very attractive to big corporations). Even with the low rate, the franchise tax generates 14 percent of the state's general fund revenues—Delaware collected over $126 million in 1986.
6. The corporate tax rate itself is a low 8.7 percent and is collected only on money generated inside Delaware. Compare this to the 10 percent New York State tax and the total burden of 19 percent for companies operating within New York City.

Favorable Corporation Law

1. Delaware's court of Chancery sets the nation's standards for sophistication and timeliness in shaping corporate law. Don-

ald E. Schwartz, professor of law at Georgetown Law Center, says: "There is, by an order of several magnitudes, a larger body of case law from Delaware than there is from any other jurisdiction, enabling not only lawyers who practice in Delaware, but lawyers everywhere who counsel Delaware corporations to be able to render opinions with some confidence." By quickly establishing precedents on the issues that confront corporate heads today, Delaware has defined the legal parameters for doing business faster and more comprehensively than any other state. Business leaders feel more secure in making decisions and planning for the future, because the law is set early; as Schwartz puts it, "Corporate managers and their lawyers seek predictability."

2. In Delaware, only a majority of shareholders of a company need agree to incorporate a company. Many states require a two-thirds majority.

3. Delaware allows mergers to proceed with less intrusion than just about any other state.

4. Once incorporated, a corporation can change its purpose of business without red tape from the state.

5. The corporation's terms of existence is perpetual in Delaware. Some states require renewals, which involve paper work and extra expense.

Favorable Treatment of Corporate Leaders

Delaware has recently enacted several laws designed to make life easier for corporate heads, particularly boards of directors.

1. Delaware law allows corporations to indemnify directors, officers, and agents against expenses and often against judgments, fines, and costs of settlements incurred in suits against them filed by third parties.

2. Delaware law makes it difficult to unseat directors of a corporation.

3. Directors of a Delaware corporation do not necessarily have to meet in Delaware. Decisions can be made by conference

call; they can even take an action without any meeting if there is unanimous written consent.

4. Perhaps most important in this category, Delaware passed an enabling act that allowed corporations to limit or eliminate outside directors' personal financial liability for violations of their fiduciary duty (including potential liability for gross negligence). This rule makes it much easier to attract directors to Delaware corporations; would-be directors in many states are forced to pay high liability insurance premiums to protect themselves against just such lawsuits. Although Delaware law does not allow directors to escape unscathed for perpetrating fraud, the knowledge that they won't be held up for making a mistake (even a "gross" one) makes directors happy to work in the state.

Other factors also make Delaware attractive to corporations. Unions are not as entrenched in Delaware as in most areas of the Mid-Atlantic and Northeast. Pay and cost-of-living scales are lower than in surrounding regions.

Perhaps the most enticing nontangible asset of Delaware in attracting business is the accessibility of government officials to business people. State Insurance Commissioner David Levinson was quoted in the *Forbes* article on this subject: "If you have a problem and you're operating a company in Delaware, within 48 hours you can have in one room the governor, the insurance commissioner, the president pro tem of the senate and the speaker of the house."

Delaware's probusiness slant has revived what was once a stagnant economy. But has this infusion of incorporations helped the average citizen of Delaware, when most companies do not relocate there? Evidence suggests that money has trickled down. Although there are pockets of poverty in Delaware, unemployment is now well below the national average.

The secret weapon of Delaware is its small size. A bigger state would need promises of a large number of jobs before offering financial concessions to corporations. But a small state like Delaware can siphon off the gravy and thrive. For example, Delaware offers some tax breaks to out-of-state banks if they incorporate in Delaware and maintain an office with at least one hundred employees. To a multinational bank, one hundred jobs is a drop in the bucket. To a state with fewer than twenty thousand unemployed people, one hundred jobs represents a substantial opportunity.

Why Does Coca-Cola From a Small Bottle Taste Better Than Coca-Cola From a Large Bottle or Can?

Scratch any Coca-Cola diehard, and you are likely to find someone who insists that Coke tastes best out of 6½-ounce glass bottles. While purists can handle the 12-ounce glass bottles, their eyes inevitably become glazed or downcast as soon as the words "plastic" or "can" or "three-liter" are bandied about.

We must confess. We have the same conviction. Yet the small Coke bottles are almost impossible to find in many parts of the country. A friend of ours, Chris Geist, who lives in Bowling Green, Ohio, once wrote a letter to Coca-Cola complaining that his family couldn't find the small bottles anywhere. A representative from Coca-Cola came to his door, one day, like Michael Anthony on *The Millionaire,* with a case of 6½-ounce glass bottles. The Coca-Cola man bewailed the demise of the small glass bottles himself, but explained that supermarkets refuse to stock them. Grocery stores don't like the breakability of glass, and they prefer the higher profit margins provided by the larger sizes.

Despite many attempts, we could not get Coca-Cola to respond to this Imponderable, on or off the record, but we consulted several other soft-drink companies and soft-drink technologists and got pretty much the same answer from all. The response of the National Soft Drink Association is typical: "What many people do not realize is that the exact same product that goes into small bottles also goes into large bottles and cans. We have found in the past that most perceived taste preferences disappear when a blind taste test is administered using a similar sample container such as a paper cup." Our guess, however, is that these comparisons were made under ideal conditions; the consumer is not always granted such favorable benefits. For one, the polyester resin material used in two- and three-liter soft drink bottles does not retain CO_2 as well as glass or aluminum. And although polyester resin provides a decent shelf life in the supermarket, the carbonation retention is not nearly as good as in other containers.

Most important, any amount of air in the head space (the top of the bottle) adversely affects the taste of the liquid. Once a bottle is opened, even if the closure is reapplied tightly, the carbonation and the taste are never quite the same again. Indeed, one of our pet theories is that simply in the time that it takes to fill a three-liter bottle, more air is trapped inside than in a small bottle. The amount of CO_2 in a bottle affects not only our sense of its carbonation but also our taste perception.

We didn't get far on the can versus bottle argument. We assumed that canned soft drinks didn't taste as good as bottled ones because traces of the metal affected the flavor. Soft-drink experts insist that in blind taste tests, consumers can't tell the difference, and buyers like the convenience and durability of the can. Bah humbug!

Submitted by Ronald C. Semone, of Washington, D.C.

Why and When Was 1982–1984 Chosen to Replace 1967 as the Base Year for the Consumer Price Index? Why Wasn't It Changed After Ten Years?

The Consumer Price Index measures the average change in prices over time in a fixed market basket of goods and services. The CPI tracks the prices of food, clothing, housing, utility costs, medical care, drugs, transportation, and many other goods and services, as well as the taxes on all of these items. Prices are collected from 85 urban areas, including about 4,000 food stores, 24,000 rental units, and 28,000 business establishments, both commercial enterprises and nonprofit institutions such as hospitals.

The Consumer Price Index is based on data compiled from

the Consumer Expenditure Surveys. The accuracy of these data is important not only in the abstract, to economists and researchers, but in the pocketbook, to workers, whose cost-of-living provisions are keyed to the CPI. The reference base year is changed to make comparisons of rates of change easier for the public to understand. By changing the base year every ten years or so, the numbers don't get unwieldy. Until 1987, the base year for the CPI was 1967; the index was expressed as 1967 = 100. If the 1972 index had increased by 50 percent since 1967, it would have been expressed as 150. An index rating of 250 would mean a 150 percent jump since the reference base year.

The Bureau of Labor Statistics is not under any statutory requirement to change its base year at any particular time, though it strives to do so approximately every decade. Originally, BLS intended to change the base year to 1977, which would have been a fortuitous choice for two reasons. First, the then most recent quinquennial (five-year) economic censuses were taken for 1977, and many economic time series are tied to economic censuses. Second, the expansion of the economy in 1977 after the recession of 1974–1975 was relatively balanced, which meant that there were few extreme conditions in particular segments of the economy that would have made 1977 unrepresentative of the recent period.

The most difficult part of compiling the CPI isn't the collection of raw data (it is time consuming, but not hard, to find out the cost of, say, dairy products in different localities throughout the country), but rather the weighting of the various sectors that the CPI measures. In order for the index to be accurate, it must measure not only how much goods and services cost the consumer, but also what percentage of consumer expenses each category represents. It matters little if the cost of maraschino cherries skyrockets if very few consumers buy them in the first place. But if the price of a staple such as oil zooms during an energy crisis, it will have a more substantial effect on the CPI.

In the late 1970s, the Bureau of Labor Statistics was in the process of developing new systems to "weight" the different components of the index, so the plan was to wait to implement 1977 as the base year until 1981, after the new weighting system was introduced. No base-year changes can be made without review by the Office of Federal Statistical Policy and Standards, Office of Management and Budget. According to Patrick Jackman, at the Office of Prices and Living Conditions of the Bureau of Labor Statistics, severe budget constraints led the BLS to ask for a postponement of the change, and they were granted it. Without these budget problems, 1977 would have been the reference base year, at least through the late 1980s. In other words, Reaganomics killed the planned base-year change.

So why have they finally made the decision to adopt the three-year 1982–1984 period as the reference base? Historically, the reference base period has tended to be three years (before 1967, 1957–1959 and 1947–1949 were the reference bases), mainly because it was believed there would be less volatility in the index if it reflected a fairly lengthy time period. Then a government interagency task force decided that the Bureau of Labor Statistics should designate a *single* year as the reference base period, and 1967 was chosen. The intent was to key several economic indexes to the same base year. The Office of Management and Budget abandoned this attempt and reverted to a three-year period—1982–1984—largely because this particular period was already being used in determining the expenditure base weights.

Submitted by Daniel Marcus, of Watertown, Massachusetts.

Why Do Firehouses Have Dalmatians?

Although today dalmatians serve primarily as firehouse mascots, back in the days of horse-drawn hose carts, they provided a valuable service. Dalmatians and horses get along swimmingly, so the dogs were easily trained to run in front of the carts and help clear a path for the firefighters to get to a fire quickly.

The breed eventually became so popular in New York City firehouses that the Westminster Kennel Club offered a special show class for dalmatians owned by New York Fire Department members. In their book, *Dalmatians: Coach Dog, Firehouse Dog*, Alfred and Esmeralda Treen tell the story of a Lieutenant Wise of the NYFD who had such a close attachment to his firehouse mascot, Bessie, that she followed him home on his days off, literally hopping on the streetcar to accompany him. If Bessie missed the car, she knew where to stand to catch the next one. Bessie also followed Lieutenant Wise into burning buildings but stayed one floor below the fighting line, "for fear a dog might cause a man to stumble if retreat was ordered."

Dalmatians have been used throughout history for serious work: as sentinels on the borders of homeland Dalmatia and Croatia, during wars; as shepherds; as draft dogs; as hound dogs; as retrievers; and as performing dogs (dalmatians not only are intelligent, but have excellent memories). Dalmatians' speed and endurance and lack of fear of horses enabled them to become superb coach dogs for fire carts. As the Dalmatian Club of America puts it, the breed was able "to coach under the rear axle, the front axle, or most difficult of all, under the pole between the leaders and the wheelers."

The death knell of the dalmatian as a coaching dog for fire departments sounded with the introduction of motorized cars. Dalmatians like Bessie lost their function, as we can see from this sentimental lament from Lieutenant Wise:

> For five and a half long years Bessie cleared the crossing at Third Avenue and Sixty-seventh Street for her company, barking a warn-

ing to surface-car motormen, truck drivers, and pedestrians, and during all that time she led the way in every one of the average of forty runs a month made by No. 39. Then like a bolt from the sky the white horses she loved were taken away, even the stalls were removed, and the next alarm found her bounding in front of a man-made thing that had no intelligence—a gasoline-driven engine. Bessie ran as far as Third Avenue, tucked her tail between her legs and returned to the engine house. Her heart was broken. She never ran to another fire.

Today's dalmatian is as likely to help in quashing fires as is Smokey the Bear. But firehouse mascots still abound, and the dalmatian is still often chosen by many firefighters in honor of its heroism in the past.

Doughnut-Shop Employees Always Pick Up the Doughnuts With a Tissue So That Their Hands Never Touch the Doughnuts. Why Do They Then Put the Same Tissue in with the Doughnuts for the Customer to Carry Home—Germs and All?

Imponderables must have the most discerning readership in the cosmos. Who else but an *Imponderables* reader would raise such an important health issue, hitherto hidden in obscurity?

We contacted all of the biggest doughnut-store companies. Those that responded sheepishly admitted that the tissue, usually Sav-R-Wrap tissue, is used by employees for sanitary reasons, but couldn't explain why the tissue, "germs and all," is stuffed in the bag, except that it is, as Carl E. Hass, president of Winchell's says, "placed on top of the donuts so that the customer may use it to remove the product from the bag." The director of product marketing at Dunkin' Donuts, Glenn Bacheller, agreed that perhaps the custom isn't the greatest idea, and said that the Dunkin' Donuts training department is looking into the possibility of not stuffing in the "used" tissue. Our reader may have earned a niche in doughnut history.

Submitted by Karen Simmons, of West Palm Beach, Florida.

Why Is Scoring Three Goals in Hockey Called a Hat-Trick?

"Hat-trick" was originally an English cricket term used to describe the tremendous feat of a bowler's taking three wickets on successive balls. The reward for this accomplishment at many cricket clubs was a new hat. Other clubs honored their heroes by "passing the hat" among fans and giving the scorer the proceeds. The term spread to other sports in which scoring is relatively infrequent—"hat-trick" is also used to describe the feat of scoring three goals in soccer.

According to Belinda Lerner, of the National Hockey League, the expression surfaced in hockey during the early 1900s: "There is some confusion about its actual meaning in hockey. Today, a 'true' hat-trick occurs when one player scores three successive goals without another goal being scored by other players in the contest."

Submitted by Ron Fishman, of Denver, Colorado.

What Are the Names on the Bottom of Grocery Sacks?

The kraft paper grocery sack celebrated its hundredth anniversary in 1986. Innovation is necessary to keep most venerable products fresh and appealing. But aren't designer grocery sacks a little much? What will we have next—generic paper sacks?

The largest manufacturer of grocery bags, Stone Container Corporation, of Huntington, Long Island, makes over 25 billion bags and sacks every year (a "sack" is technically a large grocery bag). These numbers are impressive, but let's face it: making the same paper bags day in and day out is not the most exciting job in the world. So many manufacturers have decided that, in order to promote pride and a sense of responsibility in the workers, employees would "sign" their work.

In most cases, the name you see on the bottom of the bag (usually just a first name) is that of the person who actually made the bag. Sometimes it is the name of the inspector responsible for supervising the production of the bags by other workers. While inspection certificates in clothing (such as "Inspected by No. 7") are there to monitor quality control, the names on grocery bags are intended solely to add to the esprit de corps.

At the Stone Container Corporation, many of the workers who run the bag machines have their full names on the bag, preceded by the words, "Produced with Pride by:" If most manufacturers included that inscription, this Imponderable wouldn't exist, and the mystique of "Toms," "Dicks," and "Harriets" on the bottom of shopping bags wouldn't linger.

Submitted by Kathi Sawyer Young, of Encino, California.

Why Do Chinese Restaurants Use Monosodium Glutamate? Other Ethnic Restaurants Don't Use It And It Certainly Is Not a Traditional Part of the Native Cuisine.

Actually, you are just as likely to encounter MSG in a Japanese or Indian restaurant as in a Chinese restaurant. And MSG *has* been a traditional part of Chinese cuisine. Sort of.

Glutamate has been used by Oriental cooks for more than two millennia, but not glutamate out of a shaker. Chefs noticed that soup stock made from *Laminaria japonica,* a certain seaweed, tasted better. They didn't know at the time that what was special about this strain of seaweed was that it contained large amounts of natural glutamate.

The discovery of the links between seaweed, monosodium glutamate, and flavor is credited to Professor Kikunae Ikeda of the University of Tokyo, who, in 1908, isolated the specific component of *Laminaria japonica* that enhanced flavor. While Western scientists believed there were only four basic tastes—sweet, sour, salty, and bitter—Ikeda believed in what the Japanese call *umami,* or tastiness, as a separate component. The mystery of *umami* is what the professor was trying to discover by unraveling the composition of the seaweed.

Dr. Ikeda was so enthusiastic about the power of MSG that commercial production of the substance was soon undertaken in Japan. MSG is essentially a concentrated form of sodium. It is extracted from seaweed in the Orient and from seaweed, beets, and grains in the West. Bean curd and soy sauces also tend to contain MSG. Japan and China are still the largest consumers of MSG, but it is used throughout the world (about 200,000 tons of it annually), and the United States, which didn't start producing it until the 1940s, eats its share—not only in Chinese and Japanese restaurants, but in many prepared foods.

One of the reasons many people associate MSG with Chinese food is the dreaded "Chinese restaurant syndrome," whose sufferers insist that they have strong allergic reactions to the chemical. One Chinese restaurant in New York City was so beset by requests from Occidentals to omit the ingredient from its food that all the waitresses donned white aprons with large block letters emblazoned with the immortal words, "No MSG."

The Glutamate Association (yes, there is a trade organization just for glutamate producers) insists that MSG is perfectly safe. They argue that MSG is no different from the glutamate that is liberated by our bodies when we eat food protein, and that MSG added to food represents only a small fraction of the glutamate contained naturally in most foods. For example, most recipes call for half a teaspoon of MSG per pound of meat. With these proportions, the MSG in a serving of chicken would constitute less than 10 percent of the glutamate already found in the chicken.

While the U.S. Food and Drug Administration lists MSG as "Generally Recognized as Safe," a category that includes other common food additives such as salt, pepper, and sugar, many Americans feel that any food that can cause intense allergic reactions must be harmful, and despite its advocacy even the Glutamate Association cites the World Health Organization's recommendation to limit MSG consumption to a maximum of one-third of an ounce a day (far more than the average American eats). So the question persists: why do restaurateurs and food packagers insist on including an additive that many people don't want and can't really distinguish by taste even if included?

MSG's greatest attribute is its ability to blend flavors well, especially the flavor of mixed spices, which is a reason why Indian cooks, so spice-conscious, tend to use MSG in great quantities. MSG tends to soften the astringent qualities of some foods. Tomatoes taste less acidic with MSG; potatoes less earthy; onions less strong.

For many of the same reasons, a number of chefs disdain MSG, believing that it deadens the taste of foods and is too often used to compensate for inferior products. Others feel there is a thin line between "blending" and "harmonizing" tastes and making all foods taste the same. Some object, in principle, to the idea of using any additive (even a naturally derived one) to enhance the natural taste of a meat, though this same argument could be made against using salt or pepper.

Many consumers confuse MSG with meat tenderizers (which often contain MSG as an ingredient). MSG does not act as either a spice or a tenderizer. Tenderizers act through the use of purified papain, a protein-splitting enzyme extracted from the juice of unripe papayas. It is confusion between MSG and tenderizers that is often the cause of nervous confrontations in Chinese restaurants. The waiters think that the Americans are being oversensitive to MSG, particularly because, more often than not, MSG will be contained in the base of sauces even if your "no MSG" orders are followed—your request will merely prevent additional MSG from being sprinkled on. The customer is sure that MSG is an ingredient the sole purpose of which is to reduce tough, rubbery meat into something edible, while the cook is likely to dash a few more sprinkles of MSG on his own food if he has the chance.

Submitted by Ronald C. Semone, of Washington, D.C.

Why Do Old Men Wear Their Pants Higher Than Young Men?

We couldn't find too many clothing or gerontological authorities who specialized in this subject, but just about everyone we contacted was more than happy to offer speculations. So, until the *Encyclopaedia Britannica* prints an entry on "Pants Height," we'll just have to write the definitive treatment ourselves.

The most common explanation we received from clothiers is that as men get older, they lose a little height and the spine curves a bit. Thus, a pair of pants that might have hung at waist level a few years ago will now creep upward. The one problem with this theory: when old men go to buy new pants, they still buy pants that hang high.

The second most popular explanation of this nagging social problem is the "Paunch Theory," which postulates that high-riding pants are a pathetic attempt to hide abdominal excesses. While this idea makes eminent sense, it doesn't explain why younger men with truly awesome beer bellies wear their pants low and let their stomachs hang unfettered.

But one wise man, Fred Shippee, of the American Apparel Manufacturers Association, offered the most logical explanation. Pants were cut relatively high until the jeans craze of the 1960s. Most men above the age of fifty, let alone much older men, never had the experience of wearing pants designed to hang on the hips rather than on the waist, and they feel uncomfortable starting now. To someone used to wearing high-cut pants, hip huggers feel as if they are always on the verge of falling off.

The low cut of jeans (called "low-rise") truly revolutionized menswear and literally lowered our sights. When we see "high-rise" pants, we think "old" or "nerd." While much has been written about the ups and downs of women's hemlines, we hope that this chapter will stir compassion for men who wear their pants high or low, and eventually bring about world peace.

Why Are Oreos Called "Oreos"?

Although the world's most popular cookie recently celebrated its seventy-fifth anniversary, the origin of its name is shrouded in mystery. It was first marketed in Hoboken, New Jersey, on March 6, 1912 as the Oreo Biscuit. In 1921, the name was changed to the Oreo Sandwich. In 1948, the same cookie was renamed the Oreo Creme Sandwich. Ultimately, in 1974, the immortal Oreo Chocolate Sandwich Cookie was born, a name that should last forever or until it is changed again, whichever comes first.

Of course, the lack of definitive facts has not deterred Oreo scholars from speculating on why those four magic letters were thrown together. Michael Falkowitz, a representative of Nabisco

Brands Customer Relations, offers two of the more popular theories:

1. Mr. Adolphus Green, first chairman of the National Biscuit Company (founded in 1898 from the consolidation of the American Biscuit Co., the New York Biscuit Co., and the United States Baking Co.) was fond of the classics. The name "Oreo" is Greek for "mountain." It was said in early testing that the cookie resembled a mountain.

2. The name was derived from *or*, the French word for "gold." The original Oreo label had scrollwork in gold on a pale green background, and the product name was also printed in gold.

After the first printing of this book, we received several letters from Greek readers indicating that the Greek word for "mountain" is not *oreo* but *oros*. Furthermore, a Greek word that, phonetically, sounds like *oreo*, means "nice," "attractive," and even "delicious." Could this, and not Nabisco's "mountain theory," be the real answer to the origins of our number one cookie?

Submitted by Ronald C. Semone, of Washington, D.C.

In the illustration:

Funny Numbers
© 1987

Printed in U.S.A.
10 9 8 7 6 5 4 3 2 1

Have a Nice Read!

Funny Numbers:
A Shocking Exposé of the Publishing World
by A. Digit

What Are Those Funny Numbers on the Bottom of Copyright Pages in Books?

Look at the bottom of the copyright page of this book. Unless someone at the typesetter's has played an elaborate practical joke on us, you should see a row of numbers with a few letters —something like this:

87 88 89 90 ??? 10 9 8 7 6 5 4 3 2 1

These funny numbers are simply a code designed to signify which printing of the book you have purchased. The number farthest to the right is the printing your have bought (in the example above, you would have bought the first printing). Harper & Row and some other publishers also mark the year of the printing, indicated by the number at the far left (in the example above, 1987). This practice is reassuring to paranoid writers, who assume that the shelf life of their works in bookstores will approximate that of *Tiger Beat* magazine on the newsstand. Those three question marks on our example stand for the three letters that Harper & Row uses to designate which of their printers handled the particular book.

When a book goes into a second printing, instead of redoing the whole copyright page, the printer merely deletes the "1" in the lower right corner and deletes the line that indicates that this was a first printing, if there is one. Some publishers put the lower numbers on the left and the higher ones on the right and omit the year of the printing altogether, but the principle is the same—the lowest number you see always conveys the print-run number.

The public often misunderstands the importance of printings. Ads blare, "Now in Its Fifth Printing!!!" as if a large number of printings guarantees a megaseller. Actually, multiple printings indicate that a book has surpassed expectations, but not necessarily that it has sold more than a book that never went into a second printing.

Submitted by Jonathan Sabin, of Bradenton, Florida.

Why Is 40 Percent Alcohol Called 80 Proof?

Before the nineteenth century, the technology wasn't available to measure the alcohol content of liquids accurately. The first hydrometer was invented by John Clarke in 1725 but wasn't approved by the British Parliament for official use until the end of the century. In the meantime, purveyors of spirits needed a way to determine alcohol content, and tax collectors demanded a way to ascertain exactly what their rightful share of liquor sales was.

So the British devised an ingenious, if imprecise, method. Someone figured out that gunpowder would ignite in an alcoholic liquid only if enough water was eliminated from the mix. When the proportion of alcohol to water was high enough that black gunpower would explode—this was the *proof* of the alcohol.

The British proof, established by the Cromwell Parliament, contained approximately eleven parts by volume of alcohol to ten parts water. The British proof is the equivalent of 114.2 U.S. proof. More potent potables were called "over proof" (or o.p.), and those under 114.2 U.S. proof were deemed "under proof" (or u.p.).

The British and Canadians are still saddled with this archaic method of measuring alcohol content. The United States's system makes slightly more sense. The U.S. proof is simply double the alcohol percentage volume at 60° F. For once, the French are the logical nation. They recognize the wisdom in bypassing "proof" and simply stating the percentage of alcohol on spirits labels. The French method has spread to wine bottles everywhere, but hard liquor, true to its gunpowder roots, won't give up the "proof."

Submitted by Robert J. Abrams, of Boston, Massachusetts.

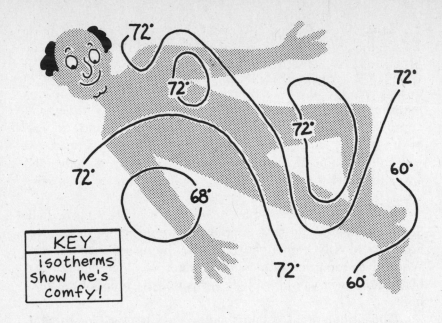

72°
72°
72°
72°
72°
72°
68°
60°
72°
60°

KEY
isotherms
show he's
comfy!

Why Are Humans Most Comfortable at 72° F? Why Not at 98.6° F?

We feel most comfortable when we maintain our body temperature, so why don't we feel most comfortable when it is 98.6° F in the ambient air? We would—if we were nudists.

But most of us cling to the habit of wearing clothes. Clothing helps us retain body heat, some of which must be dissipated in order for us to feel comfortable in warm environments. Uncovered parts of our body usually radiate enough heat to meet the ambient air temperature halfway. If we are fully clothed at 72° F, the uncovered hands, ears, and face will radiate only a small portion of our heat, but enough to make us feel comfortable. Nude at 72° F, we would feel cold, for our bodies would give off too much heat.

Humidity and wind also affect our comfort level. The more humid the air, the greater ability it has to absorb heat. Wind can also wreak havoc with our comfort level. It hastens the flow of the heat we radiate and then constantly moves the air away and allows slightly cooler air to replace it.

Submitted by Joel Kuni, of Kirkland, Washington.

Why Do They Call Large Trucks "Semis"? Semi-*Whats?*

The power unit of commercial trucks, the part that actually pulls the load, is called the "tractor." The tractor pulls some form of trailer, either a "full trailer" or a "semitrailer." According to Neill Darmstadter, senior safety engineer for the American Trucking Associations, "A semitrailer is legally defined as a vehicle designed so that a portion of its weight rests on a towing vehicle. This distinguishes it from a full trailer on which the entire load, except for a drawbar, rests on its own wheels."

Semi is short for "tractor-semitrailer," but most truckers use the term *semi* to refer to both the trailer alone and the tractor-semitrailer combination. Since the tractor assumes part of the burden of carrying the weight of the semitrailer, the "semi" must have a mechanism for propping up the trailer when the power vehicle is disengaged. The semitrailer is supported by the rear wheels in back and by a small pair of wheels, called the landing gear, which can be raised and lowered by the driver. The landing gear is located at the front of the semitrailer, usually just behind the rear wheels of the tractor.

Submitted by Doug Watkins, Jr., of Hayward, California.

Why Does San Francisco Sourdough Bread Taste Different from Other Sourdough French Breads?

This Imponderable has been a source of controversy for a long, long time. San Francisco sourdough French bread has exactly the same ingredients as any other: flour, salt, water, and natural yeast. Yet, somehow, it *is* different: the crust is dark and hard, but crumbly, and the taste more sour than the competition's.

The usual explanation for the unique San Francisco taste is the Pacific Ocean air, the winds, and especially the fog. Many of the old-time bakers still use brick ovens, and a mystique has built up around them.

The answer, however, can actually be traced to the "starter." Whenever a baker makes a batch of sourdough, some of the fermented dough is set aside as a starter, which will be used to leaven the next batch. In this manner, the action of the yeasts is maintained continuously. Some of the older bakeries have strains of starters dating back more than one hundred years.

A tourist at Fishermen's Wharf is literally eating a descendant of the bread consumed by forty-niners.

The use of starters did not originate in the United States. It is believed that the Egyptians, four thousand years before the birth of Christ, were exposing dough to airborne yeast spores to ferment. Until the advent of commerical yeasts and baking powders, most breads were leavened by using leftover dough from previous batches.

The majority of San Francisco bakeries use a proportion of about 15 percent starter. Little yeast is needed, because the starter contains natural yeast. Sourdough bread made with chemical yeast tends to lack the sourness and tang of the bread made with starter.

Scientists have also tried to pierce the mystique of San Francisco sourdough. In 1970, Dr. Leo Kline and microbiologist colleagues at Oregon State University isolated two organisms in sourdough: *Saccharomyces exiguus,* an acid-tolerant yeast; and a rod-shaped bacterium, resembling lactic-acid bacteria, which they suggested be named *Lactobacillus san francisco.* According to trade newspaper *Milling and Baking News,*

> The newly-discovered lactic acid bacteria causes souring. Lactic acids and acetic acids are produced by fermentation of carbohydrates in the flour and provide the sour flavor while the acetic acid primarily keeps spoilage and disease-producing bacteria from growing in the dough.
>
> Yeast cells do the leavening; carbon dioxide gas is produced by yeast during fermentation of carbohydrates in flour. Ethyl alcohol, also produced by the yeast cells, evaporates during cooking. The carbon dioxide provides the light, fluffy texture important in bread, biscuits and pancakes.

The specific bacterium *Lactobacillus san francisco* seems to be a new species, which probably explains why San Francisco sourdough is unique.

Submitted by Donald C. Knudsen, of Oakland, California.

Does the U.S. Postal Service Add Flavoring to the Glue on Postage Stamps to Make the Taste More Palatable?

The Postal Service doesn't intend the adhesive to have any particular flavor. The glue on U.S. postage stamps comes in only two "flavors," and not for reasons of taste.

The first type, used primarily on commemorative stamps, is simply a blend of corn dextrin (a gummy substance extracted from starch) and water. This solution is gentle on commemoratives, which are designed to last longer (in philatelic collections) than "regular" stamps.

The second type of adhesive, used on regular issues, such as the twenty-two—cent flag stamps, is a blend of polyvinyl acetate emulsion and dextrin. Added to this scrumptious taste sensation is a bit of propylene glycol, used to reduce paper curl.

Is the taste of the stamp glue reminiscent of another flavor? Dianne V. Patterson, of the Postal Service's Consumer Advocate's Office, points out that the polyvinyl acetate used for stamp adhesive is the basic ingredient in bubble gum.

Submitted by Joel Kuni, of Kirkland, Washington.

Why Do Wagon Wheels in Westerns Appear to Be Spinning Backward?

Motion-picture film is really a series of still pictures run at the rate of twenty-four frames per second. When a wagon being photographed moves slowly, the shutter speed of the camera is capturing tiny movements of its wheel at a rate of twenty-four times per second—and the result is a disorienting strobe effect. As long as the movement of the wheel does not synchronize with the shutter speed of the camera, the movement of the wheel on film will be deceptive. This effect is identical to disco strobe lights, where dancers will appear to be jerking frenetically or listlessly pacing through sludge, depending on the speed of the strobe.

E. J. Blasko, of the Motion Picture and Audiovisual Products Division of Eastman Kodak, explains how the strobe effect works in movies: "As the wheels travel at a slower rate they will appear to go backward, but as the wheel goes faster it will then become synchronized with the film rate of the camera and appear to stay in one spot, and then again at a certain speed the wheel will appear to have its spokes traveling forward, but not at the same rate of speed as the vehicle." This strobe effect is often seen without need of film. Watch a roulette wheel or fan slow down, and you will see the rotation appear to reverse.

Submitted by Richard Dowdy, of La Costa, California. Thanks also to: Thomas Cunningham, of Pittsburgh, Pennsylvania, and Curtis Kelly, of Chicago, Illinois.

Why Does Unscented Hair Spray Smell?

Let's differentiate between the major formulations of hair spray. A "regular, scented" formula contains fragrance added to give the spray a distinctive smell. "Fragrance free" refers to hair sprays without any fragrance added at all. Very few hair sprays are fragrance free; these products are designed to be hypo-allergenic for consumers sensitive to any fragrance.

"Unscented" hair spray not only *does* have a scent, but it also has added fragrance, a fragrance designed to mask the chemical base of the product. The difference between "unscented" and "scented" hair spray, then, is that "scented" hair spray contains more fragrance, more long-lasting fragrance, and a fragrance designed to be prominent.

Hair spray is 95 percent alcohol. John Corbett, of the Clairol Corporation, told *Imponderables* that neat alcohol doesn't smell like a dry martini, but rather like the rubbing alcohol used for cleaning cassette recorder heads. The prospect of spraying rubbing alcohol on your hair is an affront to your nose. The purpose of the unscented spray is to avoid challenging the more expensive, and desirable, fragrance of perfumes or colognes that the customer might apply.

Corbett added that he doubted that consumers cared much about how hair sprays smelled, as long as they weren't offensive. In the toiletry category, consumers seem to care passionately only about the smell of shampoos.

Sales for "scented" and "unscented" hair sprays are about even, with "fragrance free" products representing a tiny share of the market.

Submitted by Aleta Moorhouse, of Mesa, Arizona.

Why Does Heat Lightning Always Seem Far Away? And Why Don't You Ever Hear Thunder During Heat Lightning?

Heat lightning is actually distant lightning produced by an electrical storm too far away to be seen by the observer. What you see is actually the diffused reflection of the distant lightning on clouds.

You don't hear thunder because the actual lightning is too far away from you for the sound to be audible. There *is* thunder where the lightning is actually occurring.

Why Can't Hair Grow on a Vaccination Mark?

A vaccination mark is nothing more than scar tissue. A vaccination causes an inflammation intense enough to destroy the hair follicles in its vicinity. Any deep injury to the skin will destroy hair follicles and cause hair loss, a condition known to dermatologists as "scarring alopecia." One can easily transplant hair onto a vaccination mark, if desired, but one can never bring a dead hair follicle back to life.

Submitted by David Wilsterman, of Belmont, California.

What Kind of Hen Lays Extra-Large Eggs? What Determines the Size Categories of Chicken Eggs?

Although there are six official sizes of eggs, the smallest size a grocery store consumer is likely to encounter is the medium egg. Sizes are determined strictly by weight, as the chart below indicates:

Size	Minimum Weight Per Dozen Eggs
Jumbo	30 oz.
Extra-Large	27 oz.
Large	24 oz.
Medium	21 oz.
Small	18 oz.
Peewee	15 oz.

Small- and peewee-sized eggs are generally sold to bakers and food processors at a lower price per pound than larger eggs, so the prudent egg producer wants to encourage hens to lay big eggs. The Single-Comb White Leghorn, the most popular laying hen in the United States, eats approximately a quarter-pound of feed per day. It takes about four pounds of feed to produce a dozen eggs, so larger eggs are not without cost to the farmer.

The biggest variable in egg size is the age of the chicken. Generally speaking, the older the chicken, the larger the egg. Hens that start laying eggs prematurely tend to lay more but smaller eggs. Different breeds also tend to vary in size of eggs produced. Leghorns, for example, tend to lay larger than average eggs.

The weight of the bird is another factor in egg size. A pullet (a hen less than one year old) significantly underweight at sexual maturity will tend to produce small eggs. For this reason, farmers must pay attention not only to the quantity but the quality of feed given to hens. Feed without sufficient protein and fatty acids, while cheaper to supply, will yield smaller eggs. Hard evidence suggests that hatching environment also affects egg size. Heat, stress, and overcrowding all lower the size of eggs.

Consumers are often confused about the relative value of different sizes of eggs. Some feel that larger sizes have disproportionately more shell than smaller eggs (not true: shells constitute approximately 10 percent of the weight of all eggs). Which size will constitute the best buy is likely to vary from week to week and can be determined by a formula devised by the American Egg Board. Let's say large eggs cost 96 cents a dozen and a dozen extra-large eggs cost $1.05. Which is the better buy? First, find the price difference by subtracting the price of the smaller size from that of the larger. In this case, the price difference is $1.05 minus 96 cents, or 9 cents. Then divide the price of the smaller eggs by 8 to find the "magic number." In this case, 96 cents divided by 8 is exactly 12 (round off the number if it isn't even).

If the magic number is lower than the price difference, the

smaller eggs are a better buy. If the magic number is higher than the price difference, the larger eggs are a better buy. Because in the example 9 (the price difference) is less than 12 (the magic number), the extra-large eggs would be cheaper per weight than the large eggs.

You divide the price difference by 8 to find the magic number because egg sizes vary in increments of exactly 3 ounces. Large eggs (the one size that is always available in stores) are 24 ounces per dozen, so 24 divided by 3 equals 8, which when divided into the price of a dozen eggs provides the benchmark price for 3 ounces of eggs.

All of this makes perfect sense, even though we sense eyes glazing over as you read it.

Submitted by Helen M. Tvorik, of Mayfield Heights, Ohio.

Why Do Some Chickens Lay Brown Eggs and Others Lay White Eggs?

The color of eggs comes exclusively from the pigment in the outer layer of the shell and may range from an almost pure white to a deep brown, with many shades in between. The only determinant of egg color is the breed of the chicken.

Because white eggs are preferred in almost every region of the country, the Single-Comb White Leghorn has become by far the favorite egg-layer in the United States. The Leghorn is prized for many reasons: it reaches maturity earlier than most pullets; it utilizes its feed efficiently; it is relatively small (an important consideration when most chickens are kept in cages even smaller than New York City studio apartments); it is hardy, adapting well to different climates; and most important, it produces a large number of eggs. If more consumers went along with New England's preference for brown eggs, more breeds such as the Rhode Island Red, New Hampshire, and Plymouth Rock would be provided to produce them.

A simple test to determine the color of a hen's eggs is to look at her earlobes. If the earlobes are white, the hen will lay white eggs. If the earlobes are red, she will produce brown eggs.

Although many people are literally afraid to try brown eggs, they are no more or less nutritious or healthy than white ones. In fact, brown eggs have some cachet among health-food aficionados, which guarantees their higher cost, if not greater benefits.

Egg yolks also range dramatically in color, but yolk variations are caused by dietary differences rather than genetic ones. Yolk color is influenced primarily by the pigments in the chicken feed. If the hen gets plenty of yellow-orange plant pigments known as xanthophylls, the pigments will be deposited in the yolk. Hens receiving mash with yellow corn and alfalfa meal will lay eggs with medium yellow yolks. Those fed on wheat or barley produce lighter yolks. A totally colorless diet, such as white corn, will yield a colorless yolk. For cosmetic reasons alone, farmers avoid giving chickens a colorless diet, because consumers prefer a yellowish hue to their yolks.

Submitted by Jo Ellen Flynn, of Canyon Country, California.

How Did the Expression "Two Bits" Come to Mean 25 Cents? How Did "Two-Bit" Come to Mean "Cheap"?

"Bit," which has long been English slang for any coin of a low denomination, derived from the Old English word, *bite*, which meant a small bit or morsel. Before the American Revolution, English money was in short supply, so coins from all over Europe, Mexico, and South America were equally redeemable. Sailors and new immigrants assured a steady stream of non-English coins into the new country. Because there were so many different denominations, coins were valued by their weight and silver and gold content.

Spanish and Mexican coins were especially popular in early America. "Bit" became a synonym for the Spanish and Mexican coin, the *real*. The real was equivalent to one-eighth of a peso, or twelve and one-half cents. Particularly in the southwestern United States, where the Mexican influence was most strongly felt, Americans rarely called a quarter anything else but "two bits." Recognizing that U.S. coinage had no equivalent to one bit, Southwesterners usually referred to ten cents as a "short-bit" and fifteen cents as a "long-bit," and occasionally still do so today.

In Spain, a bit was an actual coin. Pesos were manufactured so that they could literally be cut apart. A peso, which equaled eight bits, could be cut in half to become two four-bit pieces. Cut in fourths, a peso became four two-bit pieces.

How did the term "two-bit" become synonymous with cheapness and tackiness (especially because, obviously, one bit is cheaper than two)? The first known use of this meaning, according to word whiz Stuart Flexner, was in 1856, referring to a saloon that was so cheap that a good, stiff drink could be had for . . . two bits.

Submitted by John A. Bush, of St. Louis, Missouri. Thanks also to: Tom and Marcia Bova, of Rochester, New York.

Why Did Volkswagen Discontinue Making the "Bug"?

At the time, most of us thought that the phasing out of VW Beetles was an insidious plot, designed to eliminate a model that lasted too long and thus reaped insufficient profits for the greedy automaker. Why try to sell us a cheap, dependable Bug, when they could peddle a more expensive, less charismatic Rabbit? But there was a simpler explanation: the VW Beetle had stopped selling.

Although the German Bug had been around since the 1930s, the first one didn't hit North American shores until 1949, when two brothers brought them home and soon decided to open a Volkswagen distributorship on the East Coast.

Five years later, only 8,000 Beetles were sold in a year, but the Bug soon started to take off. In 1955, sales reached 32,000 units. The peak years for the Bug were the mid-1960s: in 1968, Volkswagen sold 423,000 automobiles, the vast majority of them Bugs. But watch what happened to Beetle sales after 1968:

1969:	403,000
1970:	405,000
1971:	354,000
1972:	358,000
1973:	371,000
1974:	243,000
1975:	92,000

After a slow, but steady decline from its peak year, the popularity of the Beetle fell precipitously. Volkswagen was forced to bail out of the Bug business in the United States.

Speculations about the metaphysical reasons for the decline of the Beetle abound. Everything from growing materialism to the Vietnam War has been blamed for its demise. Larry Brown, a representative of Volkswagen, offers a more prosaic excuse: Japanese imports. In the late 1960s, the Japanese began to provide a stylish subcompact with better specifications than the Beetle, and for less money.

During its heyday, the VW Bug was designed to attract the thinking man or woman. The Beetle appealed to college professors and students, not only because it was cheap, but because it was emblematic of their rationality and refusal to buy into the car-as-fantasy-machine myth that Detroit had been perpetuating for decades. The irony, of course, is that the VW became a status symbol itself—the ultimate antiestablishment automobile. VW ads delighted in tweaking the pretensions of more expensive cars. The Beetle buyer felt confident that he was buying the steak, and not the sizzle.

Offered a rational alternative, these same buyers flocked to the Japanese imports and later to Volkswagen's own Rabbit, which offered better mileage, better handling, better leg and shoulder room, and a safer ride. The Beetle was superior only in its charm, in its charisma.

The charisma remains. Brown told *Imponderables* that he still receives calls from longtime owners of Beetles, asking him to come for a thirtieth birthday party or a 300,000-mile party. Magazines and clubs devoted to the Beetle keep the spirit alive. The Beetle is still manufactured in Mexico, and *exported* to Germany.

We asked Brown if the Beetle might be resurrected. After all, the death of convertibles was prematurely announced just a few years ago. Brown wasn't optimistic. He estimates that there are twenty thousand hardcore addicts who would buy a new Beetle immediately, but after that, it would be a hard sell. Volkswagen would have to sell a Beetle with the old specifications for approximately eight to ten thousand dollars today, overpriced for an entry-level car.

Instead, for 1988 Volkswagen has chosen to offer the Fox, a new entry-level model with a list price of under six thousand dollars. Intended to compete with the new Korean cars, as well as Japanese subcompacts, the Fox won't be the cheapest car on the block, the way the old Beetle used to be.

The race of so many full-line automobile companies to compete in the entry-level field indicates that there is still profit to be made in low-priced cars. Volkswagen probably would have clung to the Beetle indefinitely as long as it sold, but the Bug's demise, ironically, forced Volkswagen to expand its thinking and planning. A one-product company (the Karmann Ghia, VW's sports car, was never a big seller in the U.S.) is always in jeopardy.

The success of the Beetle was dependent on the serendipitous confluence of several factors: a bulging demographic group of baby boomers coming into driving age; a clever marketing and advertising campaign; and a growing wave of antiestablishment thinking. But most of all, there was that charisma. When charisma wanes, it's hard to regain that magic. Ask the makers of Flavor Straws. Or Screaming Yellow Zonkers.

Submitted by J. Spring, of Citrus Heights, California.

Why Are the Flush Handles on Toilets on the Left Side?

Have we finally found a product that was designed with the left-hander in mind? Of course not.

Most early flush toilets were operated by a chain above the tank that had to be pulled down by hand. Almost all of the chains were located on the left side of the toilet, for the user had more leverage when pulling with the right hand while seated.

When the smaller handles near the top of the tank were popularized in the 1940s and 1950s, many were fitted onto existing toilets then equipped with pull-chains. Therefore, it was cheaper and more convenient to place the new handles where they fitted standard plumbing and fixtures.

The handles offered the user a new dilemma: should one flush while seated or flush while standing? Although this subject is not often discussed in polite quarters, we are more that delighted to tread on delicate matters in order to stamp out Imponderability wherever we find it. Alexander Kira, in his wonderful book, *The Bathroom*, notes that in the "Cornell Survey of Personal Hygiene Attitudes and Practices in 1000 Middle-Class Households," 34 percent of respondents flushed the toilet while still seated and 66 percent flushed while standing up. Thus, it would seem that the majority of Americans flush either left-handed or else in an awkward right-handed crossover style. Would there be reason to switch handles over to the right side?

In *The Bathroom*, Kira argues that the current configuration discriminates not so much against right-handers as against flushing-while-seated types:

> Most flushing mechanisms are poorly located. . . . convenient only if the user flushes the closet after rising and turning around. A sizable number of persons prefer, however, for one reason or another (odor, peace of mind, and so on), to flush the closet while seated and after each bowel movement and must engage in contortions to do so. Since the water closet is presently also used for

standing male urination, this might be regarded as a justification for its location.

Kira sees the flushometer as no solution to our left-right problem. Generally used only in public bathrooms, flushometers are those levers that you never know whether you are supposed to operate with your foot or your hand. Evidently, people use both, making the flushometer unsanitary. The device's position, about eighteen inches off the floor, is awkward for either extremity.

Europeans have fared little better in tackling this design problem. Most European toilets have a pull-up knob located on top of the tank. The placement of the knob not only makes it most difficult to flush from a seated position, but it prevents using the top of the tank as a magazine rack or radio stand.

Alexander Kira's solution to all of these problems is Solomonlike in its ecumenicalism. He recommends a spring-loaded flush button set into the floor that would allow users to flush from either a seated or standing position, "before, during, or after elimination." These buttons can be operated electronically rather than mechanically, freeing them from the fate of the current flush handle, the placement of which is dictated by the demands of mechanics rather than the convenience of the user.

Submitted by Lisa R. Bell, of Atlanta, Georgia. Thanks also to: Linda Kaminski, of Park Ridge, Illinois.

Why Does the Price of Gas End in Nine-Tenths of a Cent?

No one we contacted in the oil or service-station businesses could find any reason to believe that gas isn't priced at $1.19.9 for the same reason that automobiles are priced at $9,999 or record albums at $8.98. As Ralph Bombardiere, the executive director of the New York State Association of Service Stations, Inc., put it, "There is and will always be a big difference between the price of 29.9 cents and 30.0 cents, and the same principle will follow through when the number reaches $1.29.9 and $1.30.0."

It is doubtful that sophisticated marketing surveys were ever undertaken by service stations or oil companies to establish the effectiveness of ending prices in nine-tenths of a cent, but the use of fractional prices goes back at least seventy years. C. F. Helvie, customer relations manager for the Mobil Oil Corporation, sent us a fascinating letter, the result of combing through Mobil's collection of photographs of old service stations and other reference materials.

Helvie found a photograph of a 1914 Texaco gas station that displayed a sign advertising gasoline for 14½ cents per gallon. The Mobil material suggests, but does not conclusively prove, that the practice of ending unit pricing of gas with nine-tenths started no earlier than the late 1920s and early 1930s.

The Great Depression decimated the demand for gasoline. More than 2.6 million cars and trucks were taken off the road, and the consumption of gasoline was down a billion gallons per year in both 1932 and 1933. Gas stations fought to survive. Helvie writes:

> Production at the time was running far above demand and the market quickly went into a serious oversupply situation. It was at that time that premiums such as candy, cigarettes, ash trays, dolls, and countless other giveaway items made their appearance at service stations. In such a competitive climate, it seems reasonable to assume that the gasoline marketers of the day would have been

attracted to the concept of fractional pricing. In addition, mechanical gasoline pumps, with computers that could be set to fractional prices, began to make their appearance at service stations at about the same time.

When prices zoomed at the gas pumps during the oil crisis of the 1970s and federal price and supply controls were imposed, individual stations lost the autonomy to set prices. The government issued mandated formulas for pricing, which resulted in unusual fractional amounts. Further compounding the problem was that, during this period, the price of gasoline went above one dollar per gallon, and most computers were incapable of handling prices of more than two digits. Until their computers could be modified, many service stations simply set their pumps to calculate half-gallon prices, which led to more strange fractions. Some stations chose to sell at a price per liter and maintained the usual nine-tenths fraction.

Consumers are accustomed to most retail establishments charging a cent or two less than a round number. Helvie indicated that his experience as a customer-relations expert was that "most motorists accept and understand gasoline prices ending in nine-tenths of a cent per gallon, but they react negatively to prices ending in other fractions."

Submitted by John D. Wright, of Hazelwood, Missouri. Thanks also to: Charles F. Myers, of Los Altos, California.

When I Open the Hot-Water Tap, Why Does the Sound of the Running Water Change As It Gets Hot?

The whistling sound you hear occurs with cold water as well, but is more common with hot water. Whistling occurs when there is a restriction of water flow in the pipes. According to Tom Higham, executive director of the International Association of Plumbing and Mechanical Officials, the source of the noise depends on the construction of the plumbing: "If the piping is copper, the cause is usually attributed to undersized piping. If the pipe is galvanized steel, noise is usually caused by a buildup of lime which reduces the area for the flow of the water." Water flow is restricted more often with hot water, as Richard W. Church, president of the Plumbing Manufacturers Institute, explains it, because "of additional air in the hot water formed when the molecules expand during the heating process."

The crackling noise you hear in the water heater is caused by lime accumulations in its tank. As the water heater expands and contracts, depending on the temperature, the lime breaks off and falls to the bottom of the tank. The water pipes simply transmit and amplify the glorious sound.

Submitted by Glenn Worthman, of Palo Alto, California.

The Measurement of "One Foot" Was Meant to Approximate the Length of a Man's Foot. How Did They Decide How Long a Meter Should Be?

The U.S. Constitution gives the Congress the power to fix uniform standards for weights and measures. Previously, little uniformity existed among different colonies or even among different countries in Europe or Asia. For example, King Henry I personally provided the nose and thumb that set the standard for the length of a yard, while other nations didn't even use the yard as a measure. Asian nations must have wondered if our "feet" really measured the length of a human foot.

Much of the clamor for a uniform system of measurement came from France. In 1790, during the French Revolution, the National Assembly of France asked the French Academy of Sciences to provide an invariable standard for all weights and measures. One committee responded quickly, urging that the Academy accept a decimal system as the simplest and most elegant solution. A subsequent committee recommended that the basic measure of length of such a system should represent a portion of the earth's circumference: a unit equal to one ten-millionth of the length of a quadrant of the earth's meridian (in other words, one ten-millionth of an arc representing the distance between the Equator and the North Pole).

This unit was later given the name *mètre*, from the Greek word *metron*, meaning "a measure." The meter was the foundation for all of the other measures, as Valerie Antoine, executive director of the U.S. Metric Association, Inc., explains:

> The unit of mass was to be derived by cubing some part of this length unit and filling it with water [thus, the "gram" became the mass of one cubic centimeter of water at its temperature of maximum density]. The same technique would also provide the capacity measure. In this way, the standards of length, mass, and capacity were all to be derived from a single measurement, infinitely reproducible because of natural origins, precisely interrelated, and decimally based for convenience.

The "metric system" did not catch on beyond France, at first, but its rigidity and standardization made it appealing to scientists and engineers throughout the world. Few people realize that as early as 1866, by Act of Congress, it was made "lawful throughout the United States of America to employ the weights and measures of the metric system in all contracts, dealing or court procedures." By the turn of the twentieth century, the supremacy of the metric system was assured among developed nations.

The advances in precision instruments made the original definition of the meter too fuzzy. The "Treaty of the Meter," an

1875 agreement, established a mechanism to refine and amend the metric system, and seventeen nations, including the United States, joined the "Metric Convention." Since 1893, the meter has been defined as the length of the path traveled by light in a vacuum during a time interval of 1/299,792,458 of a second (in other words, the speed of light in a vacuum is 299,792,458 meters per second). As the speed of light is unlikely to change in the near future, scientists are confident that the meter will have a long life as a standard measurement.

Valerie Antoine mildly reprimanded us for using the spelling "meter," which is an Americanized version of what most of the world—including other English-speaking countries—spells "metre."

Why Does the Moon Appear Bigger at the Horizon Than Up in the Sky?

This Imponderable has been floating around the cosmos for eons and has long been discussed by astronomers, who call it the moon illusion. Not only the moon but the sun appears much larger at the horizon than up in the sky. And constellations, as they ascend in the sky, appear smaller and smaller. Obviously, none of these bodies actually changes size or shape, so why do they *seem* to grow and shrink?

Although there is not total unanimity on the subject, astronomers, for the most part, are satisfied that three explanations answer this Imponderable. In descending order of importance, they are:

1. As Alan MacRobert of *Sky* & *Telescope* magazine states it, "The sky itself appears more distant near the horizon than high overhead." In his recent article in *Astronomy* magazine, "Learning the Sky by Degrees," Jim Loudon explains, "Apparently, we perceive the sky not as half a sphere but as half an oblate [flattened at the poles] spheroid—in other words, the sky overhead seems closer to the observer than the horizon. A celestial object that is perceived as 'projected' onto this distorted sky bowl seems bigger at the horizon." Why? Because the object appears to occupy just as much space at the seemingly faraway horizon as it does in the supposedly closer sky.

2. When reference points are available in the foreground, distant objects appear bigger. If you see the moon rising through the trees, the moon will appear immense, because your brain is unconsciously comparing the size of the object in the foreground (the tree limbs) with the moon in the background. When you see the moon up in the sky, it is set against tiny stars in the background.

Artists often play with distorting perception by moving peripheral objects closer to the foreground. Peter Boyce, of the American Astronomical Society, adds that reference points tend to distort perception most when they are close to us and when the size of the reference points is well known to the observer. We *know* how large a tree limb is, but our mind plays tricks on us when we try to determine the size of heavenly objects. Loudon states that eleven full moons would fit between the pointer stars of the Big Dipper, a fact we could never determine with our naked eyes alone.

3. The moon illusion may be partially explained by the refraction of our atmosphere magnifying the image. But even the

astronomers who mentioned the refraction theory indicated that it could explain only some of the distortion.

A few skeptics, no doubt the same folks who insist that the world is flat and that no astronaut has ever really landed on the moon, believe that the moon really *is* larger at the horizon than when up in the sky. If you want to squelch these skeptics, here are a few counterarguments that the astronomers suggested.

1. Take photos of the moon or sun at the horizon and up in the sky. The bodies will appear to be the same size.

2. "Cover" the moon with a fingertip. Unless your nails grow at an alarming rate, you should be able to cover the moon just as easily whether it is high or low.

3. Best of all, if you want proof of how easy it is to skew your perception of size, bend over and look at the moon upside down through your legs. When we are faced with a new vantage point, all reference points and size comparisons are upset, and we realize how much we rely upon experience, rather than our sensory organs, to judge distances and size.

We do, however, suggest that this physically challenging and potentially embarrassing scientific procedure be done in wide-open spaces and with the supervision of a parent or guardian. *Imponderables* cannot be held responsible for the physical or emotional well-being of those in search of astronomical truths.

Submitted by Patrick Chambers, of Grandview, Missouri.

If We See Mockingbirds During the Day and Hear Them at Night, When Do They Sleep?

At night, but off and on, and with an occasional nap during the day. Birds aren't as compulsive as humans are about their sleep hours, but then they don't have nine-to-five jobs. Birds also require much less sleep than humans, but then they don't have taxes to worry about either.

Actually, it has proved to be quite a challenge to determine the sleep patterns of birds. Laboratory experiments can't replicate the conditions they face in the elements, and any movement or sound the scientist makes during close observation will disrupt the sleep he is trying to measure.

No one has actually proved that sleep is physiologically necessary for birds. Its main benefit for them might be that standing still helps conserve energy: if a bird can't hear a potential predator hovering or see a worm ripe for the picking, it can't do anything about it. While sleep leaves them more vulnerable to predators, it is all that keeps birds from an exhausting 24-hour-a-day hunt for food.

Simply standing motionless with their eyes closed or open provides rest for most birds. Birds that live in the Arctic or Antarctic regions and have to contend with periods of twenty-four–hour sunlight, often take short catnaps throughout the day and night, but require no long sessions of sleep.

For diurnal birds, like mockingbirds, the daytime is full of activity, but they don't sleep peacefully throughout the dark hours, as anyone who has ever attempted to sleep near a male mockingbird knows all too well.

Submitted by Kathi Sawyer Young, of Encino, California.

Why Were Phillips Screws and Screwdrivers Developed?

The straight-bladed screwdriver was popular long before the advent of the Phillips. Was the Phillips merely a marketing ploy to make old hardware obsolete?

Fred A. Curry, a retiree of Stanley Works and now an educational consultant, has a large collection of Stanley tools and old catalogs. While trying to find an answer to our query, Mr. Curry found a 1938 article in Stanley's *Tool Talks*, which, to use a hardware metaphor, bangs the nail on the head:

> The most recent major improvement in screw design is the Phillips recessed head, self-centering screw and bolt. This type of screw is already extensively used in many of the major industries, and is even replacing the common wood screw for home repairs. Stanley has the No. 1 license to manufacture the screwdrivers, hand and

power driven bits required by the Phillips screw, and now offers a complete line of these Stanley quality drivers and bits.

The main selling point of the Phillips was clearly the self-centering feature. Straight-bladed screwdrivers tended to slip out of the screws' slots, ruining wood or other material, occasionally even injuring the worker. The recessed Phillips screws allowed a closer and tighter fit than the conventional slots. It may be harder, initially, to insert the Phillips screwdriver, but once it is in place, the Phillips is much less likely to slip.

Why Do Trucks Now Say Their Contents Are "Flammable" When They Used to Say "Inflammable"?

The prefix "in" usually means "not." If you are *in*sensitive, you are not sensitive. If you are *in*coherent, you are not coherent. If you are *in*flammable, you are not flammable.

Oops! You *are* flammable if you are inflammable.

The English language is less than a logical construct. "Flammable" and "inflammable" have identical meanings: "easily set on fire." So why did the trucking industry bother to change its warning notices?

Fire-insurance underwriters are usually given credit for starting the changeovers. They felt that foreigners, unaware of this exception to the usual meaning of "in-," might misconstrue "inflammable" signs, so they lobbied to change labels on containers and tanks to "flammable." Scientists, always sensitive to the need for international understanding, have also adopted "flammable."

Ironically, although the purpose of the change from "inflammable" to "flammable" was to facilitate the understanding of nonnative speakers, almost all of the international agencies responsible for regulating the labeling of (in)flammable materials, such as the United Nations, have chosen "inflammable"as their standard. A. N. Glick, president of the Conference on the Safe Transportation of Hazardous Articles (COSTHA), told *Imponderables* that the International Maritime Dangerous Goods Code of the International Maritime Organization uses the term "inflammable" but permits the use of "flammable" if there is a footnote reference.

The *Harper Dictionary of Contemporary Usage* had its panel of language experts (a group so concerned with preserving the English language that they still don't quite trust Edwin Newman) vote on whether they used "flammable." Most didn't, but they couldn't work up much enthusiasm for trying to fight its use, as it is less ambiguous to nonnatives.

I am surprised that nobody bothered to ask what an intelligent foreigner might think about a country in which companies bothered to put signs on their trucks announcing that the truck was carrying cargo that was *not* easily set on fire.

Submitted by Warrine Ahlgreen, of Tallmadge, Ohio. Thanks also to: Allen Johnson, Ph.D., of Kennewick, Washington.

Why Can't They Make Newspapers That Don't Smudge?

Reading a newspaper might be good for the mind, but it ain't great for the hands. After a bout with the Sunday paper, your hands are likely to look as if they have been engaged in a mud-wrestling contest rather than an intellectual endeavor.

What is that junk all over your hands? It is ink. And as much as these smudges annoy you, they bother the people within the newspaper industry even more. As Ralph E. Eary, who is responsible for the production and engineering of Scripps Howard's newspapers, told *Imponderables*, "Ink rub-off has been my mortal enemy for forty years. I have experimented with various inks, dyes, and water-based inks over the past twenty-two years and each comes up a failure."

Black news inks have changed little over the past forty

years. Inks consist of pigments, which produce colors, and "vehicles," liquids that carry the pigments. Conventional newspaper inks have an oil base. Oil never dries completely, which is why these inks smear on your hands and clothes. Black inks usually contain between 10 percent and 18 percent carbon black pigment content, with the balance consisting of mineral oil similar to automobile lubricating oil. Inks designed for letterpress machines have less pigment than ink used for offset presses.

Much hope was held out for the durability of water-based inks, but they have not proved to be a solution. In an article about ink for the journal of the American Newspaper Publishers Association, *Presstime*, technical writer Paul Kruglinski states the newspapers' continuing dilemma: "Ink rub-off is a relative problem: Its cause and elimination are not dependent on any one variable. The incidence and amount of rub-off hinge on the ingredients in inks, the kinds of inks used in each printing process and the type of newsprint used. It takes more than just changing chemicals to eliminate rub-off, researchers have learned."

Two factors have exacerbated the rub-off problem in recent years. The first is the changeover, by many newspapers, from letterpress to offset presses. In the letterpress process, the relief plate literally imprints the ink into the paper. The offset process works by what is called a "kiss" or "touch" impression, in which ink is deposited on the surface of the page, where it is more likely to smear.

The second and perhaps more significant trend over the past few decades has been toward publishers using heavier ink (adding extra pigment and oil to a particular area of page space) to make the paper more easily readable. The *New York Times*, for example, is extremely dark; the *Wall Street Journal* is printed with much lighter ink. Unfortunately, the *Times* and other newspapers pay a price for their high contrast—higher rub-off and higher "show-through" (the tendency of the print on the back side of a page to be visible on the front).

Newspaper publishers and ink manufacturers fight over who

is responsible for ink rub-off. The publishers blame the ink manufacturers for providing low-quality ink. The ink manufacturers insist that if the newspapers were willing to pay for better-quality ink, they would be glad to provide it.

The issue, clearly, is money. Now that most cities are monopolized by one paper, or by two papers owned by the same company, readers are literally a captive audience. It isn't clear to the newspapers that reduced rub-off would lead to increased sales. According to the American Newspaper Publishers Association, ink constitutes less than one percent of operating costs for most newspapers that don't publish in color.

Rub-resistant inks *are* more expensive. They work by neutralizing the carbon black in conventional inks by means of additives, such as resins and waxes. Resins trap the carbon black particles, making them stick to the surface of the newsprint. Wax works to cut down smearing by lubricating the surface of the page, reducing the friction between the ink and the fingers. The more resin and wax added to ink, the more rub-resistant it is— and the more expensive it is.

Conventional black ink designed for the letterpress process costs newspapers about thirty cents a pound; offset ink costs about fifteen cents more per pound. Most rub-resistant inks add at least ten cents more per pound to the bill. If these additives totally eliminated rub-offs, most newspapers would probably buy them, but as of now, they only improve the situation. The industry is still looking for rub-off–free ink.

Is there any solution to the rub-off problem? Ralph Eary, of Scripps Howard, and many other printers think the answer will probably come with flexography presses, which use a water-based ink. Eary believes that when the current generation of presses needs to be replaced, most publishers will choose flexo presses. Adds *Presstime*'s Paul Kruglinski,

Letterpress and offset inks are said to "dry" through the dispersion of the vehicle into the newsprint. They actually don't dry; the fibers absorb the oil. But because the vehicle in flexo inks is water, there is evaporation. Not only that, the latex additives bind the pigments in flexo inks to the surface of newsprint. With flexography, newspapers may be able to use a thinner newsprint stock for their products without quality degradation.

And we newspaper fanatics won't have to wear gloves to carry our treasures home.

Submitted by Jeff Charles, of St. Paul, Minnesota. Thanks also to: Cassandra Sherrill, of Granite Hills, North Carolina.

How and Why Do Horses Sleep Standing Up?

Horses have a unique system of interlocking ligaments and bones in their legs, which serves as a sling to suspend their body weight without strain while their muscles are completely relaxed. Thus, horses don't have to exert any energy consciously to remain standing—their legs are locked in the proper position during sleep.

Most horses do most of their sleeping while standing, but patterns differ. Veterinarians we spoke to said it was not unusual for horses to stand continuously for as long as a month, or more. Because horses are heavy but have relatively fragile bones, lying in one position for a long time can cause muscle cramps.

While one can only speculate about why the horse's body evolved in this fashion, most experts believe that wild horses slept while standing for defensive purposes. Wayne O. Kester, D.V.M., executive director of the American Association of Equine Practitioners, told us that in the wild, the horse's chief means of protection and escape from predators was its speed. "They were much less vulnerable while standing and much less apt to be caught by surprise than when lying down."

Submitted by Carole Rathouz, of Mehlville, Minnesota.

Why Is Seawater Blue and Tap Water Clear? Why Does the Color of the Ocean Range from Blue to Red?

White light consists of all the primary and secondary colors in the spectrum. Each color is distinguished by the degree to which it scatters and absorbs light. When sunlight hits seawater, part of it is absorbed while the rest is scattered in all directions after colliding with water molecules.

When sunlight hits clear water, red and infrared light absorb rapidly, and blue the least easily. According to Curtiss O. Davis of the California Institute of Technology's Jet Propulsion Laboratory, "only blue-green light can be transmitted into, scattered, and then transmitted back out of the water without being absorbed." By the time the light has reached ten fathoms deep, most of the red has been absorbed.

Why doesn't tap water appear blue? Curtiss continues: "To see this blue effect, the water must be on the order of ten feet deep or deeper. In a glass there is not enough water to absorb much light, not even the red; consequently, the water appears clear."

Thus if clear water is of a depth of more than ten feet, it is likely to appear blue in the sunlight. So how can we explain green and red oceans?

Both are the result not of the optical qualities of sunlight but of the presence of assorted gook in the water itself. A green sea is a combination of the natural blue color with yellow substances in the ocean—humic acids, suspended debris, and living organisms. Red water (usually in coastal areas) is created by an abundance of algae or plankton near the surface of the water. In open waters, comparatively free from debris and the environmental effect of humans, the ocean usually appears to be blue.

Submitted by Jim Albert, of Cary, North Carolina.

Why Don't Kitchen Sinks Have an Overflow Mechanism?

That little hole on the inside near the top of your bathroom sink or that little doohickey near your bathtub faucet is known in the plumbing trade as the "overflow." Its sole purpose is to prevent unnecessary spills when forgetful users leave water flowing unattended. Most bathtubs and bathroom sinks have such safety features, but we have never encountered a kitchen sink that did. Is there a logical reason?

Yep. Three, at least.

1. Most kitchen sinks, especially in homes, are actually double sinks. The divider between the double sinks is markedly lower than the level that would cause an overflow. Thus, excess

water in one of the sinks is automatically routed to the other side.

2. The kitchen sink is less likely than bathroom basins to go unattended for long periods of time. Because it takes so long to fill a bathtub, many a potential bather has answered the telephone, reached out and touched someone, and found much to his consternation that overflow mechanisms in bathtubs are far from infallible.

3. Perhaps the most important reason: kitchen sinks are usually made out of hard cast-iron surfaces, which tend to accumulate germs and fats more easily than china bathtubs, for example. Most kitchen overflows become quickly clogged, not only defeating the purpose of overflows, but creating unsanitary conditions.

Robert Seaman, the retired marketing manager of American Standard, told *Imponderables* that there is a current movement in the plumbing industry away from putting overflows into bathroom sinks. Germs can breed and spread inside overflows, and most get clogged eventually anyway. Many localities, however, have code requirements that mandate overflows in all lavatory sinks, where they are likely to remain until these codes are relaxed.

Submitted by Merrill Perlman, of New York, New York.

Why Do You Have to *Dry*-Clean *Rain*coats?

Actually, the majority of raincoats are washable. If the label indicates that a raincoat must be dry-cleaned, one or more components or fabrics of the coat are not washable. The most common offenders: linings (especially acetate linings), buttons, most wools, pile, satins, rubber, and canvas.

Most laymen assume that the care label instructions for rainwear refer to the effect of cleaning on water repellency. Actually, the water-resistant chemicals with which raincoats are treated are partially removed by both washing and dry-cleaning. Strangely, washing is easier on water repellency than dry-cleaning, as long as the detergent is completely removed through extra rinse cycles. According to Londontown Corp., makers of London Fog raincoats, the "worst enemies of water-repellent fabrics are (in this order) soil, detergents, and solvents." Dirt damages water repellency far more than cleaning, and stains tend to stick to raincoats if not eliminated right away.

Some of the solvents that dry cleaners use are destructive to water repellency. Before the original energy crisis, most dry-cleaning solvents were oil-based and were relatively benign to raincoats. When the price of oil-based solvents soared, the dry-cleaning industry turned to the synthetic perchloroethylene, which can contaminate water-repellent fabrics. Michael Hubsmith, of London Fog, said that if dry cleaners would rerinse garments in a clear solvent after dry-cleaning, the problem would go away. Likewise, if dry cleaners used clean dry-cleaning solution every time they treated a new batch of clothes, raincoats would retain their water repellency. But dry cleaners are as likely to blow the money for new solvent for every load as a greasy spoon is to use new oil for every batch of french fries.

Fred Shippee, of the American Apparel Manufacturers Association, adds that for many garments, clothing manufacturers have a choice of recommending either or both cleaning methods. Shippee speculates that some manufacturers might tend to favor dry-cleaning over washing for reasons of appearance. A washed raincoat needs touching up. A dry-cleaned, pressed raincoat looks great. When people like the way their garments look, they are likely to buy the same brand again.

What Is the Purpose of the White Half-Moons on the Bases of Our Fingernails and Toenails? And Why Don't They Grow Out with the Nails?

Those white moons are called lunulae. The lunula is the only visible portion of the nail matrix, which produces the nail itself. The matrix (and the lunulae) never moves, but new nails continually push forward, away from the matrix.

Why does a lunula appear white? Dermatologist Harry Arnold explains:

> The nail beds distal to the lunulae look pink because capillaries with blood in them immediately underlie the nail plate. The lunulae look white because the thin, modified epidermis of the nail bed is three or four times thicker there, being the busy factory where nail plate is manufactured. The lunula is avascular [without blood vessels], so it looks white.

> *Submitted by Joanna Parker, of Miami, Florida. Thanks also to: Jo Hadley, of Claremont, California.*

Can Raisins Be Made Out of Seeded Grapes?

At one time, no doubt, raisins had seeds. Humans have eaten raisins for at least three millennia, presumably ever since someone was hungry enough to do a little experimentation with a cluster of sun-dried grapes. We do know that raisins were a valuable commodity long before the birth of Christ, especially in the Middle East, where foods that could withstand the hot sun and store indefinitely without spoilage were prized. We know that raisins were cherished in southern Europe, as well: in ancient Rome, two jars of raisins could fetch you a slave boy in trade.

Today, raisins have no seeds. When we pop a raisin into our mouth, we are saved that moment of nervous anticipation we encounter with table grapes, wondering whether we are about to bite into a hard pip. More than 90 percent of all raisins are made from Thompson seedless grapes, exactly the same table grape that is omnipresent in produce sections of the supermarket.

When they have reached the proper ripeness, in early autumn, Thompson grapes are taken from their vines and placed on paper trays to sun-dry. It takes about two to three weeks in the sun before the raisin reaches the correct degree of moisture (15 percent, as opposed to the 78 percent water content of table grapes), and the desired color and flavor. Four to five pounds of grapes sacrifice their lives to yield one pound of raisins.

About 6 percent of the Thompson seedless crop is taken to raisin plants for immediate processing. There the grapes are cured with sulfur dioxide to preserve their color and dried in ovens. The result is golden seedless raisins (also known as "goldens"), which are popular in baking recipes, especially fruit cakes. The grapes used for golden seedless are thus identical to those that make dark brown raisins.

The tiny currants used in hot cross buns and the tart sultanas are limited in availability and used primarily by bakers. Both are seedless.

But yes, a seeded grape can be, and is, used for making raisins. Approximately one percent of the total raisin crop is derived from the seeded Muscat grape. The Muscat is also the largest and sweetest grape used to make raisins, and is therefore prized for baking. The Muscat is sun-dried on paper trays like regular Thompson seedless grapes, but it must undergo an additional step during processing. The dried Muscat raisins are puffed with steam and passed between rollers that force the seeds out. The wrinkles do a nifty job of hiding the resultant scar tissue, which is why our readers might have assumed that all raisins are made out of seedless grapes.

Actually, one other surgical maneuver is performed on all raisins during processing. They must, of course, be cleaned, but their wrinkles assure that there will be hard-to-reach crevices where dirt could hide. So raisins are first washed in tanks of hot water, which opens up the wrinkles and ensures that they have been scrubbed behind their metaphorical ears.

Submitted by Henry J. Stark, of Montgomery, New York.

When a Fly Alights on the Ceiling, Does It Perform a Loop or a Roll in Order to Get Upside Down?

The problem, as David Bodanis states it in *The Secret House*, is that "Flies, like most airplanes, lose their lift when they try to go through the air bottom-side up, and become not flies, but sinks."

We would not venture an uninformed opinion on such a weighty subject. When confronted with a fly question, we of course immediately think of contacting the Canada Biting Fly Centre (or as Maurice Chevalier preferred to call it, Centre Canadien sur les Insectes Piqueurs). Its director, Dr. M. M. Galloway, was bold enough to offer a definitive answer: "A fly lands by raising the forelegs above its head, making contact with the ceiling and then bringing its second and hind legs forward and up to the ceiling. The fly thus flips with a landing."

Bodanis points out the extraordinary efficiency of this technique: "As soon as these two front legs contact the ceiling the fly will aerobatically tuck up the rest of its body and let momentum rotate it to the ceiling. The manoeuver leaves the fly's body suspended upside down, without it ever having had to do a full roll, a remarkable piece of topological extrication."

Submitted by W. A. Nissen, of Visalia, California.

How Can "Perpetual Care" Be Assured In Cemeteries After They Run Out of Space for New Plots?

The cemetery industry has long promoted perpetual care, the notion that your burial area will be tended, well, perpetually. But how can a cemetery continue to pay the expenses for perpetual care after its source of income, new burials, is eliminated?

Stephen L. Morgan, executive vice-president of the American Cemetery Association, explained how perpetual care is supposed to work:

> By law, most private cemeteries operate as endowed care cemeteries and are statutorily required to invest a portion of the proceeds from the sale of a lot [and usually, mausoleum sales], frequently a minimum of ten percent in many states [the range is 5–30 percent], into an irrevocable trust fund. The principal of the trust cannot be spent but the trust income is used for cemetery maintenance and repairs. In this manner, income for care and maintenance will be available long after all lots have been sold. The obligation is continuing and literally perpetual, hence the term "perpetual care."

John R. Rodenburg, vice-president of the Federated Funeral Directors of America, acknowledges that the principle often breaks down in practice, "as can be seen by looking at many inner-city and country cemeteries that have fallen into disrepair." In the past, small cemeteries were frequently abandoned after they stopped generating cash flow, which is why perpetual-care laws were established in the first place.

In many towns, the responsibility for maintaining cemeteries has fallen on churches, local civic groups, and associations of property owners. Church-owned cemeteries often hand over the tending of the cemetery to a nearby for-profit funeral director.

The standards of service provided by perpetual care vary from state to state, but are almost always minimal compared to the services rendered by active cemeteries or funeral parks. In most cases, perpetual-care statutes mandate that grass must be cut and rows plowed. There is no provision, necessarily, that the grass must be leveled or the grounds landscaped with plants or flowers. Nor do most states have a regulation insisting that snow be plowed away during the winter or even that the cemetery be passable for visitors. "Perpetual care" doesn't include the maintenance of markers or memorials, either.

One small-cemetery owner we talked to said that although the interest on his perpetual-care trust was significant (more than $10,000 annually), this money still represented less than what was needed to hire one full-time employee. How can anyone be expected to maintain a cemetery properly when the income generated by the perpetual-care trust doesn't pay for one maintenance worker?

What Is the Purpose of the Pinholes Around the Sides of Screw Caps on Soft-Drink Bottles?

The sole purpose of these holes is to vent the pressure from the bottle when it is opened. As the cap is unscrewed, it is important to release this "head space" pressure as fast as possible; without these tiny holes, there would be a danger of the cap flying off. Anyone who has ever been hit by an errant champagne cork will applaud this safety feature.

Submitted by Henry J. Stark, of Montgomery, New York.

Why Are Military Medals Worn on the Left?

Military historians generally trace the custom of wearing military decorations on the left breast to the Crusaders, who wore the badge of honor over the heart. Whether this spot was chosen for its symbolic purpose or to use the badge as a shield for the heart is unclear. We do know that the Crusaders carried their shields in their left hands, freeing the right hand for manipulating a weapon. (This poses an ancient Imponderable: did left-handed Crusaders carry their shield in their right hand, exposing their heart to the enemy?)

Military decorations are a relatively recent phenomenon and were originally worn at the neck or from a sash. According to S. G. Yasnitsky, of the Orders and Medals Society of America, the practice changed in the first decades of the nineteenth century. During the Napoleonic campaigns, many awards were given to and by the different governments that participated in these wars. More and more orders were created for the lower classes, as well as medals given to all classes of the military and civil participants, with the proviso that they were to be worn "from the buttonhole."

Many fighting alliances between countries were forged during the Napoleonic period, and decorations were exchanged frequently. Medal inflation was rampant. A good soldier could expect to be decorated not only by his own country but by an ally or two as well. Buttonholes were bursting. Only tailors were happy. What could be done about this crisis?

As Yasnitsky told us:

> Common sense prevailed. No one wanted to hide his gorgeous accumulation of gold and enameled awards, so several methods were tried out. Some had their jewelers make smaller copies of these medals, so that they would all fit into one prescribed space on their uniforms. Others—and this became the more popular method—would display their own country's decoration from the buttonhole, but mount the other awards so that they extended in a line from that buttonhole, from left to right.

Why Do Bicycle Tires Go Flat When the Bike Isn't Used for Long Periods of Time?

When spring beckons, we go down to the basement, looking for our trusty, rusty bicycle, which we haven't used since autumn. More often than not, we find two flat tires. Why?

1. Air escapes from the valve stem. Although a valve stem cover will help reduce the outflow, nothing can prevent leakage completely. As K. L. Campbell, of Firestone Tire & Rubber Company, explained, "No materials are completely impervious to migration of gases (such as air) through them when there is a pressure differential between the inside and the outside. The bigger the pressure differential, the faster the migration." The typical automobile tire will lose from one-half pound to one pound of air pressure per month, even when in regular use.

2. Inner tubes of bicycles are more porous than auto tires. No inner tube can be made totally airtight. Butyl rubber, the best type of material for reducing leaks in inner tubes, is the most impermeable rubberlike substance available (car tires are made with a butyl inner liner). Less expensive, nonbutyl rubber inner tubes tend to leak even more.

3. The actual volume of air in a bicycle tire is quite small. There might be about a pint of air at sixty pounds per square inch (psi) in a typical bicycle tire, compared to five gallons of air at thirty-five psi, in an auto tire. A small loss of air volume in a bicycle tire thus affects the bike tire much more than it would affect the auto tire.

4. Bike tires typically require about twice the air pressure of car tires, making it much harder for them to maintain high air pressure.

5. Tire pressure lowers as the temperature goes down. This demonstrates Amonton's law, which postulates that for a body of ideal gas at constant temperature, the volume is inversely proportional to the pressure.

6. Bikes contain more structural hazards to tires than cars do. To quote Huffy Corporation's manager of marketing research, Robert J. Fink, "A bike wheel provides thirty-six opportunities to cause 'pinholes' via the spokes and nipples. Car wheels do not (usually) have spokes."

7. Leaks in auto tires are much less noticeable than bicycle leaks. With a naked eye or even a good kick or squeeze, we could

never detect the usual one pound per month loss of air pressure. Why not? The sheer bulk of the car tire itself, primarily, for auto tires consist of many ply layers (actually, layers of rubberized fabric surrounded by a belt topped off with tread rubber). A bicycle tire is usually one layer of tread with an inner tube. The heavier bead of the auto tire, along with its solid rims, lack of spokes, and much lower air pressure all conspire to make the contrast between a level of thirty psi and a level of thirty-five psi visually and tactilely indistinguishable.

The bike industry has tried to reduce the leakage problems by introducing plastic tires, which, though less porous, yield a stiff, shock-laden ride. Porous rubber is likely to be with us for a long, long time.

Submitted by Pat Mooney, of Inglewood, California.

How Do They Print the "M&M" on M&M's Chocolate Candies?

While doing the radio promotional blitz for the first volume of *Imponderables*, we were inundated by questions about M&M's. Don't Americans have something less fattening to worry about?

We contacted the consumer affairs division of Mars Incorporated, and although they were as helpful and friendly as could be, mere flattery, bribery, and appeals to humanitarian instincts were not sufficient to pry away a definitive answer.

Despite wild theories to the contrary, the "M&M" *is* printed on each candy by machine, but the process is proprietary. The "M&M" insignia separates the Mars product from present and future knock-offs, so the company is understandably sensitive about guarding its technological secrets. Mars did reveal that the process is similar to offset printing, from which one could infer that the stamper does not strike the sugar coating of the candy directly. Many pill manufacturers print their logos with a similar offset technique.

We might as well take this opportunity to unburden our readers of some of the other weighty M&M Imponderables.

Why Are There No Seams on M&M's?

M&M's are coated by a process called "panning." After the individual pieces of chocolate are assembled, they are placed in a revolving pan that looks like a clothes dryer. As they rotate, the chocolates are sprayed with colored sugar. Cool air is blown into the pan to harden the coating. After evaporation, an even layer of dry shell is formed. The process is repeated several times to achieve the thickness that Mars desires. No seam shows because the coating is uniform and no cutting or binding of any kind was necessary to form the shell.

What Does M&M Stand For?

Two names—Mars and Murrie, the head honchos at M&M Candies in the early 1940s.

Why Are There More Brown M&M's Than Any Other Color, and How Do They Determine the Ratio of Colors?

M&M/Mars conducts market research to answer precisely these types of questions. Consumers have shown a consistent preference for brown M&M's, so they predominate.

Why Did They Take Away Red M&M's? Why Have They Put Them Back Recently?

Red M&M's were victims of the Red Dye No. 2 scare, and were dropped in 1976. Although Mars didn't actually use Red Dye No. 2 to color the red M&M's, the company was understandably concerned that the public might be frightened. Once it decided that consumers not only would accept the red M&M's again, but would welcome them back, Mars, Inc. complied.

Although many people know that red M&M's were dropped and then brought back, few realize that the mix of colors in plain M&M's is different from the peanut version:

Color	Percent in Plain M&M's	Percent in Peanut M&M's
Brown	30	30
Yellow	20	20
Red	20	20
Orange	10	10
Green	10	20
Tan	10	0

M&M's seem to be an endless source of Imponderables. As soon as you answer one, another pops out. Why would consumers like more peanut greens than plain greens in the mix? Why would tan, the worthy companion of plain oranges and greens, be shunned completely by peanut buyers?

Submitted by Gail Kessler, of Newton, Massachusetts. Thanks also to: Marley Sims, of Van Nuys, California.

How Do Manufacturers Decide Whether Freezers Go on the Top or the Bottom of Refrigerators?

Until the early 1950s, almost all freezer compartments were top-mounted. This was a logical arrangement, for the compressor of the refrigerator, the warmest single device in the appliance, was located on the bottom, furthest away from the coldest device—the freezer. Clearly, this was the most fuel-efficient configuration, for the refrigerator unit acted to cushion the impact of the heat transference. Placing the freezer next to the compressor is a little like placing an air conditioner next to the fireplace and running them both at the same time—much energy is wasted.

Mysteriously, in the 1950s, bottom-mounted freezers became the rage. The most common rationalization for the popularity of bottom-mounts was that the freezer was used less often than the refrigerator, so it made sense to place the least used compartment in the least convenient place. A more likely explanation was that by introducing bottom-mounts, refrigerator manufacturers were able to appeal to the inherent trendiness of American consumers. Bottom-mounts made a blah appliance sexy by adding a new design element. If we bought cars with V-8 engines and huge fins that got ten miles per gallon, why not buy a refrigerator that burned electricity but showed a little panache. At their peak, bottom-mounts commanded almost 50 percent of the American refrigerator market and were clearly the premium design for those of breeding and distinction—that is, until the introduction of the side-by-side refrigerator.

Side-by-sides quickly became the choice of all upstanding, upscale Americans and have never lost that position. Today, approximately 75 percent of all refrigerators manufactured are top-mounts; 23 percent are side-by-sides; and only 2 percent are bottom-mounts. One does pay dearly for the privilege of buying a side-by-side—they are usually priced hundreds of dollars higher than their rival designs.

As might be expected, side-by-sides tend to be bought by an

older and more affluent consumer. The top-mount sells dispro-portionately more to the younger and less affluent buyer. The bottom-mount market falls between them, but sales skew toward an older clientele.

If the main argument for the top-mount is its fuel efficiency, couldn't the compressor be placed on top of bottom-mount re-frigerators? Some manufacturers do move the compressor for bottom-mounts, but there are inherent disadvantages to this scheme that counteract any energy savings. If placed near the top, a compressor would waste prime space for food storage. Also, as Blaine Keib, a spokesperson for Amana Refrigerators, told *Imponderables*, economy of scale is achieved by allowing the guts of the machinery to be identical from one model to another.

Ultimately, whether to top-mount or bottom-mount is a less than profound question to refrigerator manufacturers. Currently, we are energy conscious, so top-mounts reign supreme. A bot-tom-mount fad could revive; if so, appliance makers will be pleased to oblige.

Submitted by Steve Thompson, of La Crescenta, California.

Why Do Hot Dogs Come Ten to a Package and Hot-Dog Buns Come Eight to a Package?

In order to answer this most frequently asked Imponderable, we must acknowledge that, to some extent, this is a chicken and egg question. Officials from the hot-dog and bun industries tended to be a tad defensive about the whole issue, so let's clear the air. We aren't trying to assign blame here, only to make this world a better place to live. But to achieve this harmony, it is necessary to delve into the messy history of hot dog and hot-dog bun packaging, and to let the chips fall where they may. As the cliché goes, *somebody* has to do it.

The hot dog, of course, is simply a form of sausage, and sausages have been with us at least as far back as the ninth century B.C. (they were mentioned in Homer's *Odyssey*). We won't even go into who created the first hot dog, or where it originated, because we don't want to jeopardize *Imponderables* sales in Frankfurt-am-Main, Germany, or Wien (aka Vienna), Austria. Suffice it to say that, by the late seventeenth century, "dachshund sausages," what we now call "hot dogs," were sold commercially in Europe.

No one knows for sure who was the first person to serve a dachshund sausage in a roll, but one popular story is that a German immigrant sold dachshund sausages, along with milk rolls and sauerkraut, from a pushcart in New York City's Bowery during the 1860s. However they were consumed, dachshund sausages took New York by storm. In 1871, Charles Feltman set up the first Coney Island hot-dog stand, and Nathan's later became an institution.

It was also a New Yorker who coined the term "hot dog," in 1901. On a cold April day during baseball season, concessionaire Harry Stevens was losing his shirt trying to peddle ice cream and cold soda, so he sent his salesmen out to buy dachshund sausages and rolls. Vendors sold them to frozen customers by yelling, "They're red hot! Get your dachshund sausages while they're red hot!" Sports cartoonist Tad Dorgan, sitting in the press box bereft of ideas, drew a cartoon with barking dachshund sausages nestled warmly in their buns. Dorgan didn't know how to spell "dachshund," so he substituted "hot dog." The cartoon was a sensation, and the expression "hot dog" stuck.

The hot-dog bun, in its current configuration, was introduced at the St. Louis "Louisiana Purchase Exposition" in 1904 by Bavarian Anton Feuchtwanger. At first, he loaned out white gloves to customers to handle his hot sausages, but when the gloves weren't returned he asked his baker brother for help and was soon presented with the slotted hot-dog bun we know today.

In the early twentieth century, hot dogs were purchased not in grocery stores, but only in butcher shops. They were stored in bulk boxes, and one simply told the butcher how many "dogs" one wanted to buy. From all evidence, hot dogs then were the same size as "conventional" hot dogs are today—approximately five inches long and about 1.6 ounces in weight. Certainly, by the time the hot-dog makers automated, this size was standard.

Not until the 1940s were hot dogs sold in grocery stores in the cellophane containers we see today. Almost all of the early hot-dog companies sold hot dogs in packages of ten, making each package a convenient one pound.

Perhaps the main reason the number of buns and hot dogs per package never matched was that when hot-dog buns were first introduced, hot dogs were being sold in butcher shops in varying quantities. Sandwich rolls traditionally had been sold in packages of eight. Kaiser rolls and hamburger buns, like hot-dog buns, had always been baked in clusters of four in pans designed to hold eight rolls. This practice, more than anything else, seems to explain why hot-dog buns usually come eight to a pack.

Today, pans are manufactured to allow ten or twelve hot-dog buns to be baked simultaneously, but Pepperidge Farm, for one, told us that these pans are relatively difficult to obtain. Ekco told *Imponderables* that their eight-bun pans heavily outsell other varieties.

It is clear that the number of buns or dogs in a package is more the result of tradition than energetic planning, but certain trends are rendering this Imponderable semiobsolete. Very quietly the bun industry, and more particularly the hot-dog industry, are introducing new sizes.

Many regional hot-dog companies have long sold packages of eight wieners, often calling them "dinner franks" because their larger size makes them more appropriate to serve as an entree for dinner than as a luncheon sandwich or snack. Several companies make quarter-pounders, sold four to a package. Kosher hot dogs have traditionally been larger, and thus come with fewer dogs per package. Armour, and many other companies, are introducing even bigger frank packages (Armour sells sixteen-ounce and twenty-four–ounce packages). In the South, hot dogs are often sold in bulk two-pound bags as well as in conventional cellophane packages.

Similar innovation is entering the bakery business. Continental Baking, the largest producer of hot-dog buns (and the parent company of Wonder bread), and American Bakery now sell ten-bun packages in many areas.

None of the many companies we talked to indicated that it knew (or cared) what its compatriots in the other field were doing. American Bakeries, like Wonder, is experimenting with different packages in different regions, but not in response to what hot-dog packagers are doing. Everyone seems to want to march to his own drummer.

Imponderables humbly suggests a summit meeting at a neutral site to discuss these differences that have created chaos. Until then, we will be stuck with orphan frankfurters, left without the shelter of a bun.

Submitted by Charlie Doherty, of Northfield, Illinois. Thanks also to: Lisa Barba, of Corona, California; Tom and Marcia Bova, of Rochester, New York; Kathy A. Brookins, of Sandusky, Ohio; Sharon Michele Burke, of Menlo Park, California; Paul Funn Dunn, of Decatur, Illinois; Kent Hall, of Louisville, Kentucky; David Hartman, of New York, New York; Mary Jo Hildyard, of West Bend, Wisconsin; Mary Katinos, of Redondo Beach, California; Joanna Parker, of Miami, Florida; Mary Romanidis, of Hamilton, Ontario, Canada; Terry Rotter, of Willowick, Ohio; Glenn Worthman, of Palo Alto, California.

Frustrables

or

The Ten "Most Wanted" Imponderables

There comes a time in any writer's life when he must share with his readers his innermost torments, his doubts, his fears. We have no shame in baring our soul and admitting what has kept us from realizing our hopes and dreams: the scourge of Frustrables (i.e., Frustrating Imponderables). These are Imponderables for which we could not find a definitive answer; or those for which we could find an answer that we were almost sure was true, but could not confirm. A reward of a free copy of the next volume of *Imponderables* will be given to the first reader who can lead to the proof that solves any of these Frustrables.

FRUSTRABLE 1: *Why Do You So Often See One Shoe Lying on the Side of the Road?*

Since we initially researched this Imponderable for our first book, we have spoken to countless officials at the Department of Transportation and Federal Highway Safety Traffic Administration. We have observed that Rich Hall devoted an entire chapter to the subject in his *Vanishing America* book without really answering the question.

We have even found out that there was another soul brave enough to tackle the subject, Elaine Viets, columnist for the St. Louis *Post Dispatch*. She devoted two columns to this Frustrable. In her first column, she advanced several plausible theories:

- They are tossed out of cars during fights among kids.
- They fall out of garbage trucks.
- Both shoes in a pair are abandoned, but one rolls away.

But being a good reporter, Viets wasn't satisfied with her own conclusions. She turned to her readers, who responded with their own guesses at the causes of what some called SSS (Single Shoe Syndrome):

- Discarded newlywed shoes (you *do* see single cans on the highway, come to think of it).
- A variation on the "fighting kids theory"—they are specifically thrown out of school buses, during fights or as practical jokes.

This is the best we've been able to come up with. Anybody else have a better explanation?

FRUSTRABLE 2: *Why Are Buttons on Men's Shirts and Jackets Arranged Differently From Those on Women's Shirts?*

The party line on this Imponderable is that it stems back to the days when ladies of means were dressed by their maids. Because most people are right-handed, it is easiest for right-handers to button their clothes from left to right, the way men's buttons are now arranged. The button arrangement for women was presumably changed to make it easier for the female (ostensibly right-handed) maid to button her mistress's clothes.

A few other theories are advanced less often. One is that women usually support babies with their left arm when breast-feeding, so it was more convenient for women to breast-feed in public from the left breast. In order to shelter the baby from the cold, the theory goes, the mothers covered the baby with the right side of the dress or coat; it behooved the clothesmakers, then, to make garments for women that buttoned up from right to left.

The last theory stems from the days when men carried swords. A man needed to be ready to lunge at a moment's notice, so he kept his right hand in his coat to make sure it was warm. He could only do this if his coat opened from left to right. But why couldn't women's coats conform to the men's styles?

Frankly, these stories seem a tad lame to us. Obviously, all such explanations are obsolete. Now that so many clothes are unisex, many garment manufacturers would prefer one button styling, yet inertia guarantees the status quo will linger on.

Can anyone offer any evidence about the true origins of the button Imponderable?

FRUSTRABLE 3: *Why Do the English Drive on the Left and Just About Everybody Else on the Right?*

The explanations we have encountered trace the disparity back to everything from English versus Italian railroads to Conestoga wagons. But no proof, anywhere.

FRUSTRABLE 4: *Why Is Yawning Contagious?*

The most asked Imponderable, and we have no good answer, only a few lame theories. Who studies yawning?

FRUSTRABLE 5: *Why Do We Give Apples to Teachers?*

We haven't gotten to first base with this Imponderable.

FRUSTRABLE 6: *Why Does Looking Up at the Sun Cause Us to Sneeze?*

Is this nature's way of stopping us from staring at the sun? Does looking up expose the nostrils to floating allergens?

FRUSTRABLE 7: *Why Does the First Puff of a Cigarette Smell Better Than Subsequent Ones?*

Even the cigarette companies' research departments can't answer this one.

FRUSTRABLE 8: *Why Do Women in the United States Shave Their Armpits?*

This phenomenon makes Gillette and Schick happy, but they can't explain it.

FRUSTRABLE 9: *Why Don't You Ever See Really Tall Old People?*

Yes, we know that most people lose a few inches over their life-span, and that our population has gotten much taller since today's septuagenarians were young. But we should see a few elderly people of above-average height. Do very tall people have higher mortality rates than average-sized folks? The big insurance companies, who don't keep separate figures on death rates by height, don't seem to know. Does anyone?

FRUSTRABLE 10: *Why Do Only Older Men Seem to Have Hairy Ears?*

Endocrinologists we spoke to couldn't explain this phenomenon. Help!

Acknowledgments

This second volume of *Imponderables* was made possible by the enthusiasm and participation of the readers who bought its predecessor. In less than one year, more than five hundred people wrote with their own Imponderables, their praise, their criticisms, and their corrections. All were greatly appreciated.

Their kindness, generosity, weirdness, curiosity, enthusiasm, and sense of humor energized me. What did I do to deserve a reader like Joanna Parker, who peppered me with charming letters, and then offered to track down the answer to a knotty Imponderable herself? It is thrilling for a writer, who labors alone, to find out that he is, indeed, reaching the audience he was hoping for. I promise to read every letter that is sent to me, and to answer all that have a self-addressed stamped envelope (and, I must admit, many that do not). To all the readers of the first *Imponderables,* thank you for making this book possible.

Rick Kot was doing me favors even before I worked with him. He did me his greatest favor by becoming my editor. Rick actively pursues good food, cares about popular music, and occasionally even laughs at my jokes. Who could ask for anything more? My agent, Jim Trupin, laughs less often at my jokes, but is otherwise an invaluable friend and partner. Kas Schwan continues to produce brilliant cartoons on demand. The Atlantic Ocean is likely to dry up before Kas runs out of creative ideas. To all of my new friends at Harper & Row, who have welcomed me with enthusiasm and good humor, thank you for the support.

When I get lost in the wonderful world of Imponderability, it puts a strain on my innocent friends and family. They want to talk about the meaning of life. I want to talk about why meatloaf tastes the same in every institutional cafeteria. They want to talk about why suffering exists. I want to talk about why you forget that a hat is on your head but it still feels as if it's on after you've taken it off. The following people helped me maintain my sanity over the last year while I've wrestled with these unfathomable problems: Lori Ames; Judith Ashe; Michael Barson; Ruth Basu; Jeff Bayone; Jean Behrend; Eric Berg; Brenda Berkman; Cathy Berkman; Kent Beyer; Josephine Bishop; Sharon Bishop; Jon Blees; Bowling Green State University's Popular Culture Department; Annette Brown; Herman Brown; Alvin Cooperman; Marilyn Cooperman; Paul Dahlman; Shelly de Satnick; Linda Diamond; Diana Faust; Steve Feinberg; Fred Feldman; Gilda Feldman; Michael Feldman; Phil Feldman; Ray Feldman; Kris Fister; Linda Frank; Seth Freeman; Elizabeth Frenchman; Michele Gallery; Chris Geist; Jean Geist; Bonnie Gellas; Bea Gordon; Dan Gordon; Ken Gordon; Christal Henner; Sheila Hennes; Sophie Hennes; Uday Ivatury; Carol Jewett; Terry Johnson; Sarah Jones; Mitch Kahn; Dimi Karras; Mary Katinos; Peter Keepnews; Mark Kohut; Marvin Kurtz; Claire Labine; Randy Ladenheim; all of my friends at the Manhattan Bridge Club; Jeff McQuain; Carol Miller; Julie Mears; Phil Mears; Steve Nellisen; Debbie Nye; Tom O'Brien; Pat O'Conner; Jeanne Perkins; Merrill Perlman; Larry Prussin; Lela Rolontz; Brian Rose; Paul Rosenbaum; Tim Rostad; Leslie Rugg; Tom Rugg; Kas Schwan; Patricia Sheinwold; Susan Sherman; Carri Sorenson; Karen Stoddard; Kat Stranger; Anne Swanson; Ed Swanson; Carol Vellucci; Dan Vellucci; Julie Waxman; Roy Welland; Dennis Whelan; Devin Whelan; Heide Whelan; Lara Whelan; Jon White; Ann Whitney; Carol Williams; Maggie Wittenberg; Charlotte Zdrok; Vladimir Zdrok; and Debbie Zuckerberg.

The word about *Imponderables* got spread by radio and television talk show hosts (and their producers) and by newspaper and magazine writers. To them, my thanks, not only for help promoting the book, but for providing me with a forum for communicating directly with potential readers. Special thanks for service and graciousness beyond the call of duty to: Sally Carpenter; Rick Dees; John Gambling; Alan Handelman; Carol Hemingway; Marilu Henner; Emily Laisey; Dave Larsen; Jann Mitchell; Beth Morrison; and Tom Snyder.

Most of my time while working on this book is devoted to research, digging for answers. In a few cases, books provided vital information, but most Imponderables could be solved only with the assistance of experts. Undoubtedly, executives at Armour and Hygrade and Wonder bread and Pepperidge Farm have better things to do than to talk to me about why there are ten hot dogs in a package and eight hot-dog buns in a package, but God bless them, they did. The following people generously provided help that led directly to the solutions to the Imponderables in this book: Dennis Albert, Westwood Products; Richard B. Allen, Atlantic Offshore Fishermen's Association; Frances Altman, National Hot Dog and Sausage Council; American Council of Otolaryngology; Dr. Harold E. Amstutz, American Association of Bovine Practitioners; Gerald Andersen, Neckwear Association of America; Beth Anderson, American Institute of Baking; Valerie Antoine, U.S. Metric Association; Jan Armstrong, International Tennis Hall of Fame; Darrell Arnold, *Western Horseman*.

Glen Bacheller, Dunkin' Donuts; Bob Baker, United Lightning Protection Association; Michele Ball, National Audubon Society; Jim

Baker, WABC-TV; Richard C. Banks, American Ornithologist's Union; Dr. Pat. A. Barelli, American Rhinologic Society; Roz Barrow, Harper & Row; Rajat Basu, Citibank; H. R. Baumgardner, American Retreaders Association; Professor Don Beaty, College of San Mateo; Ira Becker, Gleason's Gym; Linda E. Belisle, General Mills; Peter Berle, National Audubon Society; E. J. Blasko, Eastman Kodak; Bob Bledsoe, Texas Instruments; Ralph Bombadiere, New York Association of Service Stations; Peter Boyce, American Astronomical Society; Dan Brigham, Visa; John J. Brill, Northeastern Retail Lumbermen's Association; Larry Brown, Volkswagen; James E. Bures, Fanny Farmer Candies; Walter F. Burghardt, Jr., American Veterinary Society of Animal Behavior; Lieutenant Colonel James A. Burkholder, U.S. Air Force Academy; Thomas F. Burns, American Spice Trade Association; Kenneth H. Burrell, American Dental Association.

George F. Cahill, National Flag Foundation; Inge Calderon, American Supply Association; Dr. Bruce Calnek, Cornell University; Doug Campbell, Northern Nut Growers Association; K. L. Campbell, Firestone Tire and Rubber Company; Joan Walsh Cassedy, Transportation Research Forum; Molly Chillinsky, Coin Laundry Association; Richard W. Church, Plumbing Manufacturers Institute; Gary M. Clayton, Professional Lawn Care Association of America; Conference on the Safe Transportation of Hazardous Articles; Charlotte H. Connelly, Whitman's Chocolates; Tom Consella, John Morrell; Dr. John Cook, Georgia Dermatology and Skin Cancer Clinic; Rhoda Cook, Montana Outfitters and Guides Association; B. F. Cooling, American Military Institute; John Corbett, Clairol Corporation; B. W. Crosby, Pepperidge Farm; Fred A. Curry, Stanley-Proto Industrial Tools.

Neill Darmstadter, American Trucking Associations; Curtiss O. Davis, Jet Propulsion Laboratory; Jack D. DeMent, Dole Fresh Fruit Company; Mrs. David Doane, Dalmation Club of America, Inc.; Joseph M. Doherty, D.D.S., American Association of Public Health Dentistry; R. H. Dowhan, GTE Products Corporation; Dr. G. H. Drumheller, International Rhinologic Society.

Ralph E. Eary, Scripps Howard; James E. Eisener, Suburban Newspapers of America; Dick Elgin, Department of Agriculture; Dr. Elliot, American Dermatological Association; Kay Englehardt, American Egg Board.

Joseph D. Fabin, Department of Transportation; Dr. John Falk; Michael Falkowitz, Nabisco Brands; Fred F. Feldman, M.D.; Debbie Feldstein, National Academy of Television Arts and Sciences; Robert J. Fink, Huffy Corporation; Lynn Flame, London Fog; Howard R. Fletcher, Muscatine Memorial Park; Bob Ford, AT&T Bell Laboratories; Edward S. Ford, D.V.M., Grayson Foundation; Lynda Frank; Dudley Frazier, New American Library; William H. Freeborn, Assistant Secretary of State, Delaware; Don French, Radio Shack; David F. Friedman, Adult Film Association of America; Marvin M. Frydenlund, Lightning Protection Institute; Frye Boots.

Scoop Gallello, International Veteran Boxers Association; Dr. M. M. Galloway, Canada Biting Fly Centre; Dr. James Gant, Jr., International Lunar Society; J. Byron Gathright, American Society of Colon and Rectal Surgeons; Mary S. Gilbert, L'eggs Products; Glutamate Association; S. Gordon, Pavey Envelope and Tag Corporation; Amey Grubbs, Dude Ranchers Association.

Gerard Hageney, American Sugar Division, Amstar; E. E. Halmes, Jr., Construction Writers Association; Darryl Hansen, Entomological Society of America; Carl E. Hass, Winchell's; Waldo Haythorne; Jim Heffernen, National Football League; C. F. Helvie, Mobil Oil Corporation; Bob Henderson, Kansas State Department of Wildlife; Tom Higham, International Association of Plumbing and Mechanical Officials; Shari Hiller, Sherwin-Williams Company; Hal Hochvert, Bantam Books; Frank Holman; Sarah K. Hood, International Banana Association; Ellen Hornbeck; Dr. Andrew Horne, Federal Aviation Authority Office of Aviation Medicine; Harry Horrocks, National Lumber and Building Materials Association; Susan R. Hubler, House Ear Institute; Michael Hubsmith, London Fog; Donald Hull, International Amateur Boxing Association; Rob Hummel, Technicolor; George Hundt, Ekco.

Patrick C. Jackman, Bureau of Labor Statistics; Beverly Jakaitis, American Dental Association; John Jay, Intercoiffure America-Canada; James H. Jensen, General Electric Lighting Group; Bob Joseph, Red Lobster.

Stanley Kalkus, Naval Historical Center; Shirlee Kalstone; Jeff Kanipe, *Astronomy Magazine;* Dr. Morley Kare, Monnell Institute; Blaine Keib, Hanna Refrigeration; Bill Keogh, American Bakeries; Michael R. Kershow, Bicycle Manufacturers of America; Wayne Kester, D.V.M., American Association of Equine Practitioners; Felix Kestenberg, Misty Harbor; Robert C. Knipe, Textile Care Allied Trades Association; John A. Kolberg, Spreckels Sugar Division, Amstar Corporation; Al Konecny, U.S. Military Academy, West Point; Rick Kot, Harper & Row; Thomas J. Kraner, American Paper Institute.

Randy Ladenheim, William Morrow and Company; Eugene C. LaFond, International Association for the Physical Sciences of the Ocean; Robert E. Lee, International Boxing Federation; Thomas A. Lehmann, American Institute of Baking; B. Leppek, Huffy Corporation; Bernard Lepper, Career Apparel Institute; Belinda Lerner, National

Hockey League; Naomi J. Linder, General Foods; Barbara Linton, National Audubon Society; Leo A. Lorenzen, American Planning Service; Joseph J. Lorfano, American Newspaper Publishers Association; Patricia Lortz, National Association of Bedding Manufacturers; Susan A. Lovin, Institute of Transportation Engineers; Robert Lute III.

Alan MacRobert, Sky Publishing Corporation; Ruth Mankin, Delaware Chamber of Commerce; Bob Manning, Londontown Corporation; James A. Marchiony, National Collegiate Athletic Association; Edward P. Marion, Professional Football Referees Association; Nikki P. Martin, Reynolds Metals Company; Doug Matyka, Georgia-Pacific Corporation; Marsha McLain, Pet Incorporated; Thomas H. McLaughlin, Western Pet Supply Association, Inc.; Robert L. Meckley, Westinghouse Electric Corporation; Jay L. Meikle, California and Hawaiian Sugar Company; B. G. Merritt, Eveready Battery Company; Elizabeth Crosby Metz; Carla Mikell, Colgate Palmolive Corporation; Rick Miller, Kansas State University; Dr. William E. Monroe, American College of Veterinary Internal Medicine; Montgomery Elevator Company; James H. Moran, Campbell Soup Company; Professor Richard Moran, Mount Holyoke College; Stephen L. Morgan, American Cemetery Association; Jane C. Mott, General Motors Technical Center; Ruth Mottram, Mars, Inc.; Thomas R. Myers, See's Candies.

National Bath Bed and Linen Association; National Weather Association; New York Public Library; H. M. Niebling, North American Wholesale Lumber Association; James W. Nixon, Whitman's Chocolates.

Richard O'Connell, Whitman's Chocolates; Larry O'Connor, American Truck Dealers; John O'Regan, CBS-TV; Carl Oppedahl.

Eleanor Pardue, Hanes Hoisery, Inc.; Dr. Lawrence Charles Parish; Dennis Patterson, Murray Bicycle Manufacturing; Dianne V. Patterson, U.S. Postal Service; Pechter Fields Bakery; Polly A. Penhale, American Society of Limnology and Oceanography; Merrill Perlman, *New York Times;* Robert J. Peterson, American National Metric Council; Diane Pindle, Hygrade Company; Jack Pollack, Keith County *News;* P. M. Preuss, Ford Motor Corporation; Dr. R. Lee Pyle, College of Veterinary Internal Medicine.

Judy Radcoff, Sherwin-Williams Company; George Rapp, Jr., University of Minnesota, Duluth; Mike Redman, National Soft Drink Association; Gloria E. Reich, American Tinnitus Association; Jim Renson, Print and Ink Manufacturers; Jeffrey Reynolds, National Dog Groomers Association; Nelson Rimensnyder, U.S. House of Representatives Committee on the District of Columbia; Eugene W. Robbins, Texas Good Roads/Transportation Association; Beverly C. Roberts, Lawn Institute;

John R. Rodenburg, Federated Funeral Directors of America; Paul Rosenbaum; Lou Rothstein, Misty Harbor; Chuck Russell, Sharp Electronics.

Donna Samelson, Sun-Diamond Growers of California; Norman J. Sanchez; Sherry Sancibrian, Texas Tech University; Starr Saphir, New York Audubon Society; Armand Schneider, Federal Express; Stanley M. Schuer, Gasoline and Automotive Service Dealers Association; Henry Schwarzchild, Capital Punishment Project, American Civil Liberties Union; Janet Seagle, U.S. Golf Association; Robert Seaman; Fred Shippee, American Apparel Manufacturers Association; Joe Skrivan, Huffy Corporation; Gary D. Smith, Heinz USA; Richard N. Smith, National Bureau of Standards; Whitney Smith, Ph.D., Flag Research Center; Pedro Sole, Ph.D., Chiquita Brands, Inc.; Dick Spencer, *Western Horseman;* Cheri Spies, Continental Baking; Terry L. Stibal; Ed Stuart, Chrysler Motors; John J. Suarez, R.P.E., National Pest Control Association; Harold Sundstrom, Dog Writers Association of America; Virgil Swanson.

Lisa M. Tate, Distilled Spirits Council of the United States; Thomas A. Tervo, Stearns & Foster Bedding Company; Sue Thompson, AT&T Bell Laboratories; Tony Lama Company; Barbara Torres, Armour Foods; Anthony P. Travisono, American Correctional Association; Jim Trupin, JET Literary Agency.

Simone van der Woude, Consulate General of the Netherlands; Vermont-American Corporation; Elaine Viets, St. Louis *Post-Dispatch.*

Jim Warters, Professional Golfers Association; Eric S. Waterman, National Erectors Association; Julien Weil, Royal Crown Cola; Belinda Baxter Welsh, Procter & Gamble; Al B. Wesolowsky, *Journal of Field Archaeology;* S. C. White, National Hardwood Lumber Association; David K. Witheford, National Research Council, Transportation Research Board; Maggie Wittenberg; Merry Wooten, Astronomical League.

About a thousand people were contacted to gather information for this book. Not everyone was cooperative, but an astonishing percentage were. To those who, for whatever reason, preferred to remain anonymous, yet still provided information, I give my thanks.

Subject Index

AC and DC, 21–22
Acre, origins of, 89
Airplanes
  ear popping and, 130–132
  feet swelling and, 31–32
Alcohol, proof of, 177
Aluminum foil, two sides of, 102
Ants, sidewalks and, 37–38
Apples, teachers and, 238
Armpits, shaving of, 239
Automobiles
  side-view mirrors, 38–39
  speed limits, 143
  speedometers, 144–145
  tire tread, 72–74
  white wall tires, 149

Baked goods, 121–122
Bananas, growth of, 81–82
Bathtubs, overflow mechanisms on, 214–215
Batteries, volume and power loss in, 76–77
Beetle, Volkswagen, elimination of, 192–194
Bell bottoms, sailor and, 84–85
Bicycles
  crossbars on, 90–91
  tires and, 224–226
Birds
  death of, 120–121
  droppings of, 83
  mockingbirds, sleep pattern of, 205
Blades, razor, hotel slots for, 113
Blue, in hair rinses, 117–118
"Blue blood," origin of term, 91
Body hair
  loss of, 6–8
  pubic hair, 146
  underarm hair, 146

Body odor, human, 92
Book staining, 93–94
Boots, fence posts and, 77–81
Boxers, sniffing and, 22–23
Buns
  hamburger bottom, 32–34
  hot dog, number of in package, 232–235
Bunnies, chocolate Easter, 116
Buttons, men's and women's clothing and, 237–238

"Caboodle," "kit" and, 15–17
Cadets, military, cap throwing and, 20–21
Calculators, key pad configurations of, 14–15
Candies, shapes of chocolate, 24–25
Capital punishment, hours of executions and, 34–36
Cemeteries
  financial strategies of owners of, 95–99
  perpetual care and, 221–222
Chocolate
  Easter bunnies, 116
  shapes of, 24–25
Cigarettes, odor of, 38
Clocks
  hand movements, 150
  numerals, 151–152
Clouds, rain and, 152
Coca-Cola, containers for, 157–159
Construction sites, pine trees and, 147–148
Consumer Price Index, 159–161
Copyright pages, strange numbers on, 175–176
Countdown leader, film and, 9
Cows, milking positions and, 128–129

Height, voice pitch and, 70
Height restrictions, fences and, 28–30
Highways, shoes found on, 236–237
Hockey
    hat trick in, 165–166
    Wayne Gretzky uniform in, 18
"Holland," versus "Netherlands," use of, 65–66
Horses, sleeping posture of, 212
Hot dog buns, number of, 232–235
Hot water, noise in pipes of, 199–200

Icy roads, use of sand and salt on, 12–13
"Inflammable," versus "flammable," use of, 207–208
Ivory soap, purity of, 46–47

"J" Street, Washington D.C. and, 71
"Jack," "John" and, 43
"Jetsam," versus "flotsam," use of, 60–61
"John," "Jack" and, 43

Ketchup, bottles of, 44–45
"Kit," "caboodle" and, 15–17

Label warnings, mattress, 1–2
Laryngitis, dogs and, 53–54
Lawns, reasons for, 47–50
Leader, film, 9
Light bulbs, three-way
    burnout, 104
    functioning of, 105
Lightning, heat, 185
Liquor, proof and, 177
Lobster, leftover, 9–10
Lumber, measurement of two by fours and, 87–88
Lunula, fingernails and, 218

M&M's, 227–229
MSG, Chinese restaurants and, 168–171
Mattress tags, warning labels on, 1–2

Measurements
    acre, 89
    meter, 200–202
Medals, location of on military uniforms, 223–224
Meter, origins of, 200–202
Microphones, press conferences and, 11–12
Mockingbirds, sleep patterns of, 205
Moon, apparent size of, 202–204

"Netherlands," versus "Holland," use of, 65–66
Newspapers
    ink smudges of, 209–212
    tearing of, 64

Ocean, color of, 213
Old men
    hairy ears and, 239
    pants height and, 171–172
"Oreo," origins of name, 173–174
Overflow mechanisms, kitchen sinks and, 214–215

Paint, homes and white, 100–102
Pants, height of old men's, 171–172
Paper bags, names on, 166–167
Paper cuts, pain and, 103–104
Paperback books, staining of, 93–94
Pay toilets, disappearance of, 25–26
Pennsylvania Department of Agriculture registration, 121–122
Pepper, white, source of, 135–136
Perpetual care, cemeteries and, 221–222
Phillips screwdriver, 206–207
Pine nuts, shelling of, 94
Pine trees, construction sites and, 147–148
Pinholes, bottle cap, 223
Pork and beans, pork in, 19
Postage stamps, taste of, 182
Potholes, causes of, 27
Press conferences, microphones in, 11–12
Pubic Hair, human, 146

Raincoats, dry-cleaning of, 216–217
Raisins
  cereal boxes and, 123
  seeded grapes and, 218–219
Razor blades, hotel slots for, 113
Refrigerators, configurations of, 230–231
Reynolds Wrap, texture of two sides of, 102
Roller skating rinks, music in, 107–108
Ruins, layers of, 138–140

Sailors, bell bottom trousers and, 84–85
Salt, treatment of icy roads by, 12–13
San Francisco, sourdough bread in, 180–181
Sand, treatment of icy roads by, 12–13
Scars, hair growth and, 186
Screwdrivers, reasons for Phillips, 206–207
"Semi," origins of term, 179
Seventy-two degrees, human comfort at, 178–179
Shaving armpits, women and, 239
Shoes found on highways, 236–237
Silverstone, 3
Sinks, overflow mechanisms on, 214–215
Skating music, roller rinks and, 107–108
Sleep, twitches during, 67
Snakes, tongues of, 106
Sneezing, looking up and, 238
Soap, Ivory, purity of, 46
Soft drinks
  containers for, 157–159
  pinholes on bottle caps of, 223
Sourdough bread, San Francisco, 180–181
Speed limit, 55 mph, 143
Speedometers, 144–145
Staining, paperback books and, 93–94

Stamps, postage, taste of, 182
Stocking runs
  direction of, 124–125
  effect of freezing upon, 125–126
Sugar, spoilage of, 85
Surgeon, color of uniforms of, 86

Teachers, apples for, 238
Teflon, 3
Telephones, touch tone keypad for, 14–15
Temperature, human comfort and, 178–179
Tennis, scoring in, 3–5
Three-way light bulbs
  burnout of, 104
  functioning of, 105
Tinnitus, causes of, 115–116
Tires
  automobile tread, disposition of, 72–74
  bicycle, 224–226
  white wall, 149
Toilets
  flush handles on, 195–196
  seat covers for, 83
Touch tone telephones, keypad configurations for, 14–15
Traffic control
  55 mph speed limit and, 143
  traffic lights and, 126–127
Traffic signs, placement of dangerous curve, 119–120
Tread, disposition of worn, 72–74
Trucks, origins of term "semi" and, 179
Twenty-one gun salute, origins of, 68–70
Twitches during sleep, 67
"Two bits," origins of term, 191–192
Two by Fours, measurement of, 87–88

Underarm hair, human, 146
Uniforms, surgeons', 86
Unscented hair spray, smell of, 184

Vaccination marks, hair growth and, 186
Voices, causes of high and low, 70
Volkswagen Beetles, elimination of, 192–194

Wagon wheels in film, movement of, 183
Warning labels, mattress tag, 1–2
Washington D.C., "J" Street in, 71
Water, color of, 213
Whips, cracking sound of, 74

White paint on homes, 100–102
White pepper, source of, 135–136
White-wall tires, thickness of, 149
Window envelopes, 111
Wisdom teeth, purpose of, 137
Wrinkles on extremities, 112

X-rated movies, XXX-rated movies versus, 141–142
"Xmas," origins of the word, 75

Yawning, contagiousness of, 238

Master Index of Imponderability

Following is a complete index of all ten Imponderables® books and *Who Put the Butter in Butterfly?* The bold number before the colon indicates the book title (see the Title Key below); the numbers that follow the colon are the page numbers. Simple as that.

Title Key

1 = Why Don't Cats Like to Swim? (*formerly published as* Imponderables)
2 = Why Do Clocks Run Clockwise?
3 = When Do Fish Sleep?
4 = Why Do Dogs Have Wet Noses?
5 = Do Penguins Have Knees?
6 = Are Lobsters Ambidextrous? (*formerly published* as When Did Wild Poodles Roam the Earth?)
7 = How Does Aspirin Find a Headache?
8 = What Are Hyenas Laughing at, Anyway?
9 = How Do Astronauts Scratch an Itch?
10 = Do Elephants Jump?
11 = Who Put the Butter in Butterfly?

"All wool and a yard wide," origins of term, **11**:2
"Allemande," **11**:31
"Alligator" shirts, **9**:297–298
Alphabet, order of, **1**:193–198
Alphabet soup
distribution of letters, **3**:118–119
outside of U.S., **10**:73
Aluminum cans, crushability of, **7**:157–159
Aluminum foil
and heat to touch, **8**:145–146
two sides of, **2**:102
Aluminum foil, on neck of champagne bottles, **4**:160–161
Ambidexterity in lobsters, **6**:3–4
Ambulances, snake emblems on, **6**:144–145; **7**:239–240
American accents of foreign singers, **4**:125–126
American singles, Kraft, milk in, **1**:247–249
"Ampersand," **11**:86
Amputees, phantom limb sensations in, **1**:73–75
Anchors, submarines and, **4**:40–41
Angel food cake and position while cooling, **7**:43–44
Animal tamers and kitchen chairs, **7**:7–11
Ants
separation from colony, **6**:44–45
sidewalks and, **2**:37–38
"Apache dance," **11**:30
Apes, hair picking of, **3**:26–27
Appendix, function of, **5**:152–153
Apples, as gifts for teachers, **2**:238; **3**:218–220
Apples and pears, discoloration of, **4**:171
Apples in roasted pigs' mouths, **10**:274–275
April 15, as due date for taxes, **5**:26–29
Aquariums, fear of fish in, **4**:16–18

Arabic numbers, origins of, **3**:16–17
Architectural pencils, grades of, **7**:73
Area codes, middle digits of, **5**:68–69; **9**:287
Armies, marching patterns of, **8**:264
Armpits, shaving of, **2**:239; **3**:226–229; **6**:249
Army and Navy, Captain rank in, **3**:48–50
Art pencils, grades of, **7**:73
Aspirin
headaches and, **7**:100–102
safety cap on 100–count bottles of, **4**:62
Astrology, different dates for signs in, **4**:27–28
Astronauts and itching, **9**:208–216
"At loggerheads," **11**:104–105
Athletics, Oakland, and elephant insignia, **6**:14–15
"Atlas," **11**:154–155
ATMs
swiping of credit cards in, **10**:138–141
swiping versus dipping of credit cards in, **10**:141–142
transaction costs of, **3**:102–103
"Attorney-at-law," **11**:103–104
Auctioneers, chanting of, **9**:201–204
Audiocassette tape on side of road, **7**:250–251
Audiocassette tapes on roadsides, **9**:300
Audiotape, versus videotape technology, **3**:136–137
Audiotape recorders
counter numbers on, **4**:148–149
play and record switches on, **5**:23–24
Automobiles
batteries and concrete floors, **10**:234–236
bright/dimmer switch position, **7**:44–46; **8**:258

Banking
    ATM charges, **3**:102–103
    hours, **3**:100–101
Barbecue grills, shape of, **10**:99–102
"Barbecue," **11**:173
Barbie, hair of, versus Ken's, **7**:4–5;
        8:259–260
Barefoot kickers in football, **4**:190–191
Bark, tree, color of, **6**:78–79
Barns, red color of, **3**:189–191
Bars
    mirrors in, **10**:14–17
    sawdust on floor of, **10**:118–120
    television sound and, **10**:12–14
Baseball
    black stripes on bats, **8**:104–106
    Candlestick Park, origins of, **9**:48–
        51
    cap buttons, **9**:171–172
    caps, green undersides of, **9**:172–
        173
    circle next to batter's box in, **3**:28
    dugout heights, **5**:14
    first basemen, ball custody of, **1**:43–
        44
    home plate sweeping, by umpires,
        8:27–31
    home plate, shape of, **5**:131
    Japanese uniforms, **10**:207–208
    "K," as symbol for strikeout, **5**:52–
        53
    pitcher's mounds, location of, **5**:181;
        9:195–198
Baseball cards
    wax on wrappers, **5**:123
    white stuff on gum, **5**:122
Basements, lack of, in southern
        houses, **4**:98
Basketball
    24–second clock in NBA, **1**:29–31
    duration of periods in, **9**:65–69
Basketballs, fake seams on, **4**:155–156
Baskin-Robbins, cost of cones versus
        cups at, **1**:133–135

"Batfowling," **11**:1–2
Bathrooms
    group visits by females to, **7**:183–
        192; **8**:237–238; **9**:277–278
    ice in urinals of, **10**:232–234
    in supermarkets, **6**:157
Bathtub drains, location of, **3**:159–160
Bathubs, overflow mechanisms on,
        2:214–215
Bats, baseball, stripes on, **8**:104–106
Batteries
    automobile, weight of, **5**:101–102
    concrete floors and, **10**:232–234
    drainage of, in cassette players,
        10:259–260
    nine-volt, shape of, **6**:104; **7**:242–
        243
    sizes of, **3**:116
    volume and power loss in, **2**:76–77
"Battle royal," **11**:67
Bazooka Joe, eye patch of, **5**:121
Beacons, police car, colors on, **7**:135–
        137
"Bead," "Draw a," origins of term,
        10:168
Beaks versus bills, birds and, **10**:3–4
Beanbag packs in electronics boxes,
        6:201
Beans, green, "French" style, **10**:125–
        126
Beards on turkeys, **3**:99
"Bears [stock market]," **11**:106–107
"Beating around the bush," **11**:1–2
Beavers, dam building and, **10**:42–
        46
"Bedlam," **11**:69
Beds, mattresses, floral graphics on,
        9:1–2
Beef, red color of, **8**:160–161
Beeps before network news on radio,
        1:166–167
Beer
    and plastic bottles, **7**:161–162
    steins, lids of, **9**:95–96

temperature in Old West, **5**:17–18

twistoff bottle caps, merits of, **4**:145

Beetle, Volkswagen, elimination of, **2**:192–194

Bell bottoms, sailors and, **2**:84–85

Bells in movie theaters, **1**:88–89

Belly dancers, amplitude of, **5**:202–203; **6**:237–239

Belts, color of, in martial arts, **9**:119–123

Ben-Gay, creator of, **6**:46–47

"Berserk," **11**:68–69

Best Foods Mayonnaise, versus Hellmann's, **1**:211–214

Beverly Hills, "Beverly" in, **8**:16–17

"Beyond the pale," **11**:3

Bias on audiotape, **4**:153–154

Bibles, courtrooms and, **3**:39–41

Bicycles
clicking noises, **5**:10–11
crossbars on, **2**:90–91
tires, **2**:224–226

Bill posting at construction sites, **8**:185; **9**:268–270

Billboards, spinning blades on, **9**:61–64

Bill-counting machines, **6**:271–272

Bills versus beaks, birds and, **10**:3–4

"Bimonthly," **11**:194

"Bingo," origin of term, **5**:202; **6**:231–233; **7**:233–234

Binoculars, adjustments for, **5**:124–125

Bird droppings, color of, **3**:241

"Birdie," **11**:139–140

Birds [see also specific types]
bills versus beaks in, **10**:3–4
coloring of, **5**:11–12
death of, **2**:120–121; **6**:268
direction of takeoff, **7**:107
droppings, color of, **2**:83; **3**:241
honking during migration, **7**:108
jet lag and, **3**:33–34
migration patterns of, **5**:147–148; **9**:91–94

sleeping habits of, **5**:72–73

telephone wires and, **3**:130–131

walking versus hopping, **3**:154

worms and, **10**:65–72

Birthday candles, trick, **4**:176

"Birthday suit," **11**:122

Biting of fingernails, reasons for, **9**:223–228

Bitings in keys, **8**:59–60

"Bitter end," **11**:72

"Biweekly," **11**:194

Black
clothing and bohemians, **8**:218–221
gondola color, **4**:86–87
judges' robes, **6**:190–192
specks in ice cream, **8**:132–133
stripes on baseball bats, **8**:104–106

"Blackmail," **11**:188·

Blacktop roads, lightening of, **5**:22–23

Blades, razor, hotel slots for, **2**:113

Bleach, flour and, **3**:63–64

Blind
money counting, **3**:152–153
wearing dark glasses, **7**:93–94

Blinking and babies, **6**:158–159

"Blockhead," **11**:80–81

Blood, color of, **4**:138

Blouses, women's, lack of sleeve length sizes in, **4**:113–114

Blue
as color for baby boys, **1**:29
as color of blueprints, **4**:102–103
as color of racquetballs, **4**:8–9
in hair rinses, **2**:117–118

"Blue blood," origin of term, **2**:91

Blue Demons and De Paul University, **8**:76

Blue jeans and orange thread, **9**:74

"Blue jeans," **11**:124

"Blue moon," **11**:12

Blue plate special, origins of, **4**:198; **5**:211–213; **6**:256–257; **7**:227–228; **8**:232–233

Blue ribbons (as first prize), **3**:57–58

"Blue streak," **11**:90

Blueprints, color of, **4**:102–103

"Blurb," **11**:35–36

Blushing and feelings of warmth, **5**:113–114

Boats, red and green lights on, **4**:152–153

"Bobbies [London officers]," **11**:108; **11**:166

"Bobby pins," **11**:167

Body hair
loss of, in humans, **2**:6–8
pubic hair, purpose of, **2**:146
underarm hair, purpose of, **2**:146

Body odor, human immunity to own, **2**:92

Boeing, numbering of jets, **4**:30–31; **5**:235; **8**:251–252; **10**:269–270

"Bogey," **11**:139–140

Boiling water during home births, **6**:114–115; **7**:247–248; **8**:254

Bombs and mushroom clouds, **9**:8–9

Book staining, of paperback book pages, **2**:93–94

Books
checkered cloth on, **7**:126–127
pagination in, **1**:141–144

Boots on ranchers' fence posts, **2**:77–81; **3**:243–245; **4**:231; **5**:247–248; **6**:268; **7**:250; **8**:253; **9**:298–299; **10**:251–253

"Booze," **11**:173

Bottle caps, beer, twistoff, merits of, **4**:145

Bottles, cola, one- and two-liter, **4**:188

Bottles, soda, holes on bottom of, **6**:187–188

Boulder, Colorado, magazine subscriptions and, **4**:33–34

Bowling, "turkey" in, **4**:48–49

Bowling pins, stripes on, **6**:181–182

Bowling shoes, rented, ugliness of, **3**:173–174

Boxer shorts, versus briefs, **6**:11–12

Boxers, sniffing and, **2**:22–23; **10**:256–258

Boxes, Japanese, yellow color of, **7**:130–131

Bracelets, migration of clasps on, **7**:180; **8**:197–201; **9**:279–281

Brain, Ten percent use of, alleged, **4**:198; **5**:210–211; **6**:254–256; **8**:232

"Brass tacks," **11**:6

"Bravo Zulu," origins of, **8**:72–73

Brazil nuts in assortments, **7**:145–147

Bread
"French" versus "Italian," **7**:163–166; **8**:261–262
kneading of, **3**:144–145
staling of, **7**:125–126
yeast in, **3**:144–145

"Break a leg," **11**:32–33

Breath, bad, upon awakening, in the morning, **4**:52

Brick shape of gold bullion, **8**:32–33

Bricks
holes in, **6**:61–62
in skyscrapers, **6**:102–103

Bridges versus roads, freezing characteristics of, **5**:193–194

Bridges, covered, purpose of, **7**:132–133; **9**:296

Briefs, versus boxer shorts, **6**:11–12

Brights/dimmer switch, position of, in automobiles, **7**:44–46

Broccoli and cans, **8**:74–76

Brominated vegetable oil in orange soda, **6**:96–97

Brushes, hair, length of handles on, **7**:38–39

Bubble gum
baseball card gum, **5**:122
Bazooka Joe's eyepatch, **5**:121
bubble-making ingredients, **5**:120
flavors of, **5**:119–120
pink color of, **3**:30–31

"Buck," **11**:107

Buckles, pilgrim hats and, **10**:47–51
Buffets, new plates at, **9**:5–6; **10**:274
Bulbs, light
    and noise when shaking, **5**:46–47
    halogen, **5**:164
    high cost of 25-watt variety, **5**:91
Bull rings, purpose of, **10**:147–148
"Bulls [stock market]," **11**:106–107
"Bunkum," **11**:69–70
Bunnies, chocolate Easter, **2**:116
Buns
    hamburger bottoms, thinnness of,
        2:32–34
    hot dog, number of in package,
        2:232–235
    versus rolls, **8**:65–66
Burger King wrappers, numbers on,
    8:112–113
Burials, position of deceased, **8**:255
Burping and babies, **10**:123–124
Buses
    entry into, **5**:24–25
    idling of engines on, **7**:150–151
    ridges on sides of, **5**:171
    seat belts and, **1**:84–85
Busy signals, fast versus slow, on tele-
    phones, **4**:182
Butter
    hardening after refrigeration, **5**:88
    sticks, lengths of, **5**:42
    versus margarine, in restaurants,
        1:32–33
Butterflies
    rain and, **4**:63–64
    sneezing and coughing in, **7**:81–82
"Butterfly," **11**:19
Buttons, men's versus women's shirts,
    2:237–238; **3**:207–209; **5**:226;
    7:223
Buzzing, bees and, **9**:57–60
"BZ," origins of expression, **8**:72–73

"Cab," **11**:63
Cable TV

channel allocation of, **9**:75–76
    volume levels of, **5**:7–8
"Caboodle," "kit" and, **2**:15–17
"Caddy," **11**:132
Cadets, military, cap throwing and,
    2:20–21
Cadillacs, ducks on, **5**:174–176
Cafeterias, popularity of, among eld-
    erly, **1**:96–101
Caffeine, leftover, from decaffeinated
    coffee, **6**:195
Cakes
    angel food and cooling position,
        7:43–44
    reaction to loud noises, **7**:41–42
    seven layer, missing layers of, **6**:80–
        81
Calculators, key pad configurations of,
    2:14–15
Calico cats, gender of, **6**:131–132
Calories, measurement of, **3**:7–8
Calves, versus cows, **3**:19–20
Cameras, black color of, **3**:185; **7**:246
Camouflage, colors in, **8**:114–115
Can openers, sharpness of blades on,
    6:176–177
Canada Dry, origins of name, **9**:182–
    183
Canadiens, Montreal, uniforms of,
    7:242
Canceled checks
    numbers on, **6**:123
    returned, numerical ordering of,
        6:165–166
    white paper attachments, **6**:124–
        125
Candles, trick birthday, relighting of,
    4:176
Candlestick Park, origins of, **9**:48–51
Candy
    Baby Ruth, origins of, **8**:84; **9**:288–
        289; **10**:264–265
    caramels versus toffee, **1**:64
    Milk Duds, shape of, **8**:81–82

Candy (*cont.*)
 Oh Henry, origins of, **8**:83
 placement in vending machines,
  9:162–164
 shapes of boxed chocolate, **2**:24–25
 wrapping of boxed chocolate,
  8:122–123
"Cannot complete your call as dialed"
 telephone message, **4**:129–130
Cans, aluminum, crushability of,
 7:157–159
"Can't hold a candle," **11**:71
Capital punishment, hours of execu-
 tion and, **2**:34–36
"Capital punishment," **11**:108
"Capitalization," **11**:108
Caps and gowns, at graduations, **6**:99–
 102
Caps, baseball
 buttons atop, **9**:171–172
 green undersides, **9**:172–173
Captain, Navy and Army rank of,
 3:48–50
Caramels, versus toffee, **1**:64
Cardboard on automobile grills,
 6:188–189
Carpenter's pencils, shape of, **7**:27;
 9:290–291
CAR-RT SORT on envelopes, **5**:78–
 79
Cars (see "Automobiles"),
Cartoon characters
 Casper the ghost, **10**:246–249
 Donald Duck, **3**:150
 Goofy, **3**:64–65; **7**:102
 Mickey Mouse, **3**:32
Cash register receipts, color of, **4**:143
Cashews, lack of shells in, **1**:177;
 5:235–236
Caskets, position of heads in, **6**:8–9
Casper the ghost, identity of, **10**:246–
 249
Cassette players, audio, battery
 drainage in, **10**:259–260

Cassette tapes, audio, on roadsides,
 7:250–251; **9**:300
Cat feces, as food for dogs, **6**:35–37
Cat food, tuna cans of, **8**:8–10
"Cat out of the bag," **11**:25
"Catercorner," **11**:197
Catnip and effect on wild cats, **6**:183–
 185
Cats
 calico, gender of, **6**:131–132
 catnip, **6**:183–185
 ear scratching, **5**:48–49
 eating posture, **6**:63–64
 hair, stickiness of, **5**:163–164
 miniature, **6**:154–155
 sight, during darkness, **1**:200–201
 swimming and, **1**:86
"Cats and dog, raining," **11**:26
"Catsup," **11**:177
Cattle guards, effectiveness of, **3**:115
"Cattycorner," **11**:197
Cavities, dogs and, **10**:277–278
CDs, Tuesday release of, **10**:108–112
Ceiling fans
 direction of blades, **9**:113–114
 dust and, **9**:111–113; **10**:263–264
Ceilings of train stations, **8**:66–67
Celery in restaurant salads, **6**:218;
 7:207–209; **8**:249
Cement
 laying of, **8**:258
 versus concrete, **9**:295
Cemeteries
 financial strategies of owners of,
  2:95–99
 perpetual care and, **2**:221–222
Ceramic tiles in tunnels, **6**:135–136;
 8:257–258
Cereal
 calorie count of, **1**:38–40
 joining of flakes in bowl, **8**:115–119
 Snap! And Rice Krispies, **8**:2–4
Chalk outlines of murder victims,
 3:11–12

Champagne
aluminum foil on neck of, **4**:160–161
name of, versus sparkling wine, **1**:232–234
Channel 1, lack of, on televisions, **4**:124; **7**:242; **10**:267–268
Chariots, Roman, flimsiness of, **10**:105–107
"Checkmate," **11**:133
Checks
approval in supermarkets, **7**:210–217; **8**:245–246
numbering scheme of, **4**:38–39; **5**:236–237
out-of-state acceptance of, **8**:120–121
white paper attachments, **6**:124–125; **7**:245
Checks, canceled
numbers on, **6**:123
returned, numerical ordering of, **6**:165–166
white paper attachments, **6**:124–125; **7**:245
Cheddar cheese, orange color of, **3**:27–28; **10**:275
Cheese
American, milk in Kraft, **1**:247–249
cheddar, orange color of, **3**:27–28; **10**:275
string, characteristics of, **3**:155
Swiss, holes in, **1**:192
Swiss, slice sizes of, **9**:142–146
Chef's hat, purpose of, **3**:66–67
Chewing gum
lasting flavor, **5**:195–196
water consumption and hardening of, **10**:236–237
wrapping of, **8**:111–112
Chewing motion in elderly people, **7**:79–80
Chianti and straw-covered bottles, **8**:33–35

Chicken
cooking time of, **1**:119–121
versus egg, **4**:128
white meat versus dark meat, **3**:53–54
"Chicken tetrazzini," **11**:153
Children, starving, and bloated stomachs, **7**:149–150
Children's reaction to gifts, **8**:184; **9**:234–237
Chime signals on airlines, **7**:6–8
Chirping of crickets, at night, **10**:54–57
Chocolate
Easter bunnies, **2**:116
shapes of, **2**:24–25
white versus brown, **5**:134–135
wrapping of boxed, **8**:122–123
Chocolate milk, consistency of, **3**:122–123
"Chops," **11**:47
Chopsticks, origins of, **4**:12–13
"Chowderhead," **11**:72
Christmas card envelopes, bands around, **6**:203–204
Christmas tree lights
burnout of, **6**:65–66
lack of purple bulbs in, **6**:185–186
purple, **9**:293; **10**:280
Cigar bands, function of, **4**:54–55
Cigarette butts, burning of, **5**:45
Cigarettes
grading, **6**:112
odor of first puff, **2**:238; **3**:223–226
spots on filters, **6**:112–113
Cigars, new fathers and, **3**:21–22
Cities, higher temperatures in, compared to outlying areas, **1**:168–169
Civil War, commemoration of, **3**:168–169
"Claptrap," **11**:73

Clasps, migration of necklace and bracelet, **7**:180; **8**:197–201; **9**:279–281

Cleansers, "industrial use" versus "household," **5**:64–65

Clef, treble, dots on, **10**:210–213

Clicking noise of turn signals, in automobiles, **6**:203

Climate, West coast versus East coast, **4**:174–175

Clocks
  clockwise movement of, **2**:150
  grandfather, **4**:178
  number 4 on, **2**:151–152
  Roman numerals, **5**:237–238
  school, backward clicking of minute hands in, **1**:178–179
  versus watches, distinctions between, **4**:77–78

Clockwise, draining, south of the Equator, **4**:124–125

"Cloud nine," **11**:97–98

Clouds
  disappearance of, **5**:154
  location of, **3**:13
  rain and darkness of, **2**:152

Clouds in tap water, **9**:126–127

"Cob/cobweb," **11**:20

Coca-Cola
  2–liter bottles, **5**:243–244
  taste of different size containers, **2**:157–159

Cockroaches
  automobiles and, **7**:34; **8**:256–257; **9**:298
  death position of, **3**:133–134; **8**:256
  reaction to light, **6**:20–21

Coffee
  bags in lavatories of airplanes, **4**:64–65
  bags versus cans, **4**:146
  electric drip versus electric perk in, **4**:35

restaurant versus home-brewed, **7**:181; **8**:221

Coffee, decaffeinated
  leftover caffeine usage, **6**:195
  orange pots in restaurants, **6**:67–69

Coffeemakers, automatic drip, cold water and, **4**:173

Coins
  lettering on U.S. pennies, **7**:5
  red paint on, **7**:117
  serrated edges of, **1**:40–41
  smooth edges of, **1**:40–41

Cola bottles, one- and two-liter; black bottom of, **4**:188

Colas, carbonation in, **1**:87–88

Cold water and automatic drip coffeemakers, **4**:173

Colds
  body aches and, **3**:104–105
  clogged nostrils and, **3**:20–21
  liquids as treatment for, **4**:131–132
  symptoms at night of, **3**:163

"Coleslaw," **11**:174

"Collins, Tom [drink]," **11**:165

Color
  blood, **4**:138
  cash register receipts, **4**:143
  wet things, **4**:139

Comic strips and capital letters, **5**:55-56

Commercials, television, loudness of, **3**:81–83

Commonwealths, versus states, **7**:119–121

Computer monitors, shape of, **7**:245

Computers
  erased files on, **6**:205–206; **7**:245
  monitors, shape of, **6**:129–131

Concrete, versus cement, **9**:295

Conserves, contents of, **6**:140–141

Construction cranes, assembling of, **6**:49–51

Construction sites
  bill posting at, **8**:185; **9**:268–270

pine trees and, **2**:147–148

soaping of retail windows, **8**:185; **9**:265–267

Consumer Price Index, changes in base year of, **2**:159–161

Containers, for rain measurement, **4**:161–163

Contemporary greeting cards, shape of, **6**:70–71

Coolers, Styrofoam, blue specks on, **10**:130–131

"Cooties," **11**:19

"Cop," **11**:108

Copper bowls and egg whites, **7**:99

Copyright pages, strange numbers on, **2**:175–176

Coriolis effect, **4**:124

Corn chips, black specks on, **10**:23–25

Corn Flakes, calorie count of, **1**:38–40

Corn silk, purpose of, **3**:120

Corn tortillas, size of, versus flour, **10**:142–145

Corn, baby, supermarkets and, **10**:186–187

Cornish game hens, identity of, **6**:10–11

"Corny," **11**:76

Corvette, 1983, **9**:137–139

Cottage cheese, origins of, **3**:59

Cotton

in medicine bottles, **3**:89–90

shrinkage of, versus wool, **6**:166–167

Cough medicine, alcohol in, **3**:166

Countdown Leader, film and, **2**:9

Counterclockwise, draining, north of the Equator, **4**:124–125

Counters on VCRs and audio recorders, **4**:148–149; **6**:272–273

Coupons

cash value, **5**:7–9

expiration date, **5**:189

versus mail-in refunds, **5**:187–188

Courtooms, bibles in, **3**:39–41

"Couth," **11**:84

Covered bridges, purpose of, **7**:132–133; **9**:296

Cow Palace, naming of, **9**:52–53

Cowboy hats, dents on, **5**:6; **6**:274

Cows

calves and, **3**:19–20

milking positions, **2**:128–129

nose rings and, **10**:147–148

sticking of tongues up nostrils of, **6**:164–165

CPR training in schools, **7**:31–33

Crabs, hermit, "bathroom habits" of, **7**:74–75

Crackling sound of fire, **4**:11

Cracks on sidewalks, **3**:176–178

Cranes, construction, assembling of, **6**:49–51

"Craps," **11**:134

Cravings, food, pregnant women and, **10**:183–185

Credit card receipts, printing of, at gas stations, **1**:173–175

Credit card slips, and phone numbers, **2**:129–130

Credit cards

and waiter tips, **7**:133–135; **8**:263

dipping versus swiping and, **10**:141–142

major versus minor, **1**:231

quick swiping of, **10**:138–141

Crickets

chapped legs and chirping, **3**:47–48; **5**:238; **6**:270

chirping at night of, **10**:54–57

"Crisscross," **11**:36–37

"Crocodile" shirts, **9**:297

Crowd estimates, police and, **1**:250–253

Crowing of roosters, **3**:3

Cruise controls, minimum speeds of, **8**:124–125

Crying at happy endings, **1**:79–80

Cue balls, dots on, in pool, **10**:237–240

Cups
drinking, width of, **6**:126
shape of paper and plastic, **9**:289

Curad bandages, wrappers of, **2**:58

Curly tails of pigs, **4**:199

Currency, color of, **3**:83–84

Cursive handwriting versus printing, teaching of, **7**:34–37

Curves on highways, **7**:121–122; **8**:258

"Cut and dried," **11**:5–6

"Cut the mustard," **11**:175

Cuts, paper, pain and, **2**:103–104

Daily doubles in *Jeopardy*, difficulty of, **1**:33–35

Dalmatians in firehouses, **2**:162–163; **6**:270–271

Dams, beavers and, **10**:42–46

Dance studios, floor location of, **5**:89–91; **7**:246–247

Dancing, men versus women, **6**:218; **8**:239–240

"Dandelion," **11**:196

"Dangerous Curve" signs, placement of, **2**:119–120

Dartboards, numbering schemes of, **9**:81–84

DC, AC and, **2**:21–22

De Paul University and "Blue Demons," **8**:76

"Dead End" signs, versus "No Outlet" signs, **4**:93

"Deaf and dumb," **11**:131–132

Decaffeinated coffee
leftover caffeine usage, **6**:195
orange pots in restaurants, **6**:67–69

"Deep-six," **11**:95–96

Deer, automobile headlights and, **6**:212–214

Deer ornaments, plastic lawn, **8**:185; **9**:262–264

Dehumidifiers, operation of, **6**:41–42

Delaware, incorporations in, **2**:153–157

Dentist offices, smell of, **2**:41

Deodorant aerosols, shaking of, **3**:178

Desert, rising cool air in, **9**:149–151

Dessert, fast-food restaurants and, **1**:218–221

Detergents, laundry
bleach in, **1**:150
package sizes of, **6**:168–169

Detroit Red Wings and octopus throwing, **9**:183–186

Diagonal measurement of TV sets, **4**:37

Dictionaries
pronunciation and, **10**:169–179
thumb notches in, **5**:167–168

Diet soft drinks
calorie constituents in, **6**:94–95
phenylalanine as ingredient in, **6**:96

Dimples
auto headlamps, **8**:56
facial, **3**:23
golf balls, **3**:45–46

Dinner knives, rounded edges of, **1**:231–232

Dinner plates
repositioning of, **8**:184
round shape of, **8**:162–164

Dirt, refilling of, in holes, **7**:48–49

Disc jockeys and lack of song identification, **7**:51–57

"Discussing Uganda," origins of term, **5**:246–247; **11**:145

Dishwashers, two compartments of, **6**:109–110

Disney cartoon characters
Donald Duck, **3**:150
Goofy, **3**:64–65; **7**:102
Mickey Mouse, **3**:32

Disposable lighters, fluid chambers of, **6**:92–93

Distilleries, liquor, during Prohibition, **9**:54–56

Ditto masters, color of, **6**:133–134

"Dixie," **11**:147–148

Dixieland music at political rallies, **5**:203; **6**:234–236

"Dixieland," **11**:147–148

DNA, identical twins and, **10**:18–20

Doctors, back tapping of, **3**:145–146

Doctors and bad penmanship, **5**:201; **6**:221–225; **7**:232–233; **8**:235

"Dog days," **11**:17–18

Dogs
  barking, laryngitis in, **2**:53–54
  black lips of, **3**:38–39
  body odor of, **2**:40
  cavities and, **10**:277–278
  circling before lying down, **2**:2–3; **5**:238–239
  crooked back legs of, **8**:126–128
  Dalmatians and firefighting, **6**:270–271
  drooling in, **6**:34–35
  eating cat feces, **6**:35–37
  eating posture of, **6**:63–64
  head tilting of, **4**:198; **5**:215–217; **6**:258–259; **8**:237
  lifting legs to urinate in, **4**:35–36
  miniature, **6**:154–155
  poodles, wild, **6**:207–209
  rear-leg wiggling, when scratched, **6**:52–53
  "sic" command, **5**:51
  sticking head out of car windows in, **4**:60–61
  wet noses, **4**:70–73

Dollar sign, origins of, **7**:103–104

Dolls, hair of, **7**:4–5

Donald Duck, brother of, **3**:150

Donkey Kong, origins of, **10**:38–40

"DONT WALK" signs, lack of apostrophes in, **6**:75–76

Doors
  double, in stores, **6**:177–180
  knobs versus handles on, **7**:148–149

opening orientation of, in buildings, **4**:167

shopping mall entrances, **6**:180–181

"THIS DOOR TO REMAIN UN-LOCKED DURING BUSINESS HOURS" signs, in stores, **6**:202

"Doozy," **11**:37

"Do-re-mi," **11**:29–30

Dots on cue balls, in pool, **10**:237–240

Double doors in stores, **6**:177–180

"Doubleheader," **11**:133

Double-jointedness, **10**:229–231

Double-yolk eggs, **3**:188–189

Doughnuts
  tissues and handling in stores, **2**:164; **5**:240
  origins of holes in, **2**:62–64

Downhill ski poles, shape of, **3**:69

Dr Pepper
  origins of name, **5**:129–130; **6**:272
  punctuation of, **8**:253–254

Drains, location of bathtub, **3**:159–160

"Draw a Bead," origins of term, **10**:168; **11**:134

Dreams, nap versus nighttime, **3**:124

Drinking glasses, "sweating" of, **9**:124–125; **10**:261

Dripless candles, whereabouts of wax in, **4**:182–183

"Driveway," **11**:64–65

Driveways, driving on, versus parkways, **4**:123

Driving, left- versus right-hand side, **2**:238; **3**:209–212; **6**:248–249; **7**:223; **8**:230–231

Drooling, dogs and, **6**:34–35

Drowsiness after meals, **6**:138–139

Drugstores, high platforms in, **8**:5–7

"Dry," as terms for wines, **5**:141–142

Dry-cleaning
  French, **3**:164–165
  garment labels and, **2**:59–60

Dry-cleaning (*cont.*)
One Hour Martinizing, **3**:28–29
raincoats and, **2**:216–217
Dryers, hand
in bathrooms, **10**:266–267
"off" switches, **8**:174–176
Ducks
lakes and ponds and, **10**:256
on Cadillacs, **5**:174–176
Duels, timing of, **5**:69
Dugouts, height of, **5**:14
"Dukes," **11**:137
"Dumb [mute]," **11**:131–132
"Dumbbells," **11**:131–132
Dust, ceiling fans and, **10**:263–264

E
as school grade, **3**:198; **4**:206–209
on eye charts, **3**:9–10
"Eagle [golf score]," **11**:139–140
Earlobes, function of, **5**:87–88
"Earmark," **11**:46
Earrings, pirates and, **9**:43–45;
10:272–273
Ears
hairy, in old men, **2**:239; **3**:231–
233; **5**:227–228
popping in airplanes, **2**:130–132
ringing, causes of, **2**:115–116
Earthworms as fish food, **3**:110–112
Easter
chocolate bunnies and, **2**:116
dates of, **4**:55–56
ham consumption at, **1**:151–152
"Easy as pie," **11**:172
Eating, effect of sleep on, **6**:138–139
"Eavesdropper," **11**:109–110
Ebert, Roger, versus Gene Siskel,
billing of, **1**:137–139
Egg, versus chicken, **4**:128
Egg whites and copper bowls, **7**:99
Eggs
color of, **2**:189–190
double-yolk, **3**:188–189

hard-boiled, discoloration of, **3**:34
meaning of grading of, **4**:136–137
sizes of, **2**:186–188
"Eggs Benedict," **11**:154
"Eighty-six," origins of term, **10**:265–
266; **11**:101–102
Elbow macaroni, shape of, **4**:28
Elderly men
pants height and, **2**:171–172
shortness of, **2**:239; **3**:229–231;
6:250
Elections, U.S.
timing of, **6**:41; **8**:260–261; **9**:291–
292
Tuesdays and, **1**:52–54; **3**:239
Electric can openers, sharpness of
blades on, **6**:176–177
Electric drip versus electric perk, in
coffee, **4**:35
Electric perk versus electric drip, in
coffee, **4**:35
Electric plug prongs
holes at end of, **5**:94–95
three prongs versus two prongs,
5:191
Electricity, AC versus DC, **2**:21–22
Electricity, static, variability in
amounts of, **4**:105–106
Elephants
disposal of remains of, **6**:196–197
jumping ability of, **10**:27–29
Oakland A's uniforms, **6**:14–15
rocking in zoos, **8**:26–27; **10**:279
Elevator doors
changing directions, **8**:169–170
holes, **8**:170–171
Elevators
overloading of, **5**:239
passenger capacity in, **1**:23–24
"Eleventh hour," **11**:99–100
Emergency Broadcasting System,
length of test, **1**:117–118
"Emergency feed" on paper towel dis-
pensers, **8**:149–150

Flu, body aches and, **3**:104–105

Fluorescent lights and plinking noises, **5**:47

Flush handles, toilets and, **2**:195–196

Flushes, loud, toilets in public restrooms and, **4**:187

Fly swatters, holes in, **4**:31–32

FM radio, odd frequency numbers of, **10**:59–60

Food cravings in pregnant women, **10**:183–185

Food labels
  "FD&C" on label, **4**:163
  lack of manufacturer street addresses, **4**:85

Football
  barefoot kickers in, **4**:190–191
  college, redshirting in, **7**:46–48
  distribution of game balls, **2**:44
  goalposts, tearing down of, **8**:213–217
  measurement of first-down yardage, **5**:128–129
  origins of "hut" in, **9**:294–295
  Pittsburgh Steelers' helmet emblems on, **7**:67–68
  shape of, **4**:79–81
  sideline population in, **10**:51–53
  tearing down of goalposts, **7**:181
  two-minute warning and, **10**:150–151
  yardage of kickers in, **3**:124–125

"Fore," origins of golf expression, **2**:34

Forewords in books, versus introductions and prefaces, **1**:72–73

Forks, switching hands to use, **3**:198

"Fortnight," **11**:194

Fraternities, Greek names of, **10**:94–98

"Freebooter," **11**:110

Freezer compartments, location of, in refrigerators, **2**:230–231

Freezers
  ice trays in, **10**:92–93

lights in, **10**:82–85

French dry cleaning, **3**:164–165

French horns, design of, **5**:110–111

"French" bread versus "Italian," **8**:261–262

"French" style green beans, **10**:125–126

"French" versus "Italian" bread, **7**:165–166

Frogs
  eye closure when swallowing, **6**:115–116
  warts and, **10**:121–123

Frogs of violins, white dots on, **4**:164–165; **9**:291

Frostbite, penguin feet and, **1**:217–218

Fruitcake, alleged popularity of, **4**:197; **5**:205–209; **6**:253; **7**:226–227; **9**:274–276

"Fry," **11**:179

Fuel gauges in automobiles, **6**:273

"Fullback," **11**:138

Full-service versus self-service, pricing of, at gas stations, **1**:203–209

Funeral homes, size of, **8**:152–155

Funerals
  burials without shoes, **7**:53–54
  depth of graves, **7**:14–15
  head position of, in caskets, **6**:8–9
  orientation of deceased, **7**:106
  perpetual care and, **2**:221–222
  small cemeteries and, **2**:95–99

"Funk," **11**:75

"G.I.," **11**:52

Gagging, hairs in mouth and, **7**:76–77

Gallons and quarts, American versus British, **4**:114–115

Game balls, distribution of football, **2**:44

Gasoline, pricing of, in tenths of a cent, **2**:197–198

Gasoline gauges, automobile, **3**:5–6

Gasoline pumps
  rubber sleeves on, **6**:197
  shut off of, **3**:125
Gasoline stations, full-service versus
    self-service, pricing of, **1**:203–
    209
Gasoline, unleaded, cost of, **1**:121–
    122
Gauge, width of railroad, **3**:157–159
Gauges, fuel, in automobiles, **6**:273
Geese, honking during migration,
    7:108
Gelatin, fruit in, **3**:149–150
"Gerrymander," **11**:111–112
"Get the sack," **11**:75
"Get your goat," **11**:21
"Getting down to brass tacks," **11**:6
Gifts, children and, **8**:184
Girdles and fat displacement, **5**:75–
    76
Glass, broken, **4**:168
Glasses, drinking
  squeakiness of, **9**:205–207
  "sweating" of, **9**:124–125; **10**:261
Glasses, wine, types of, **7**:123–125
Glitter, sidewalks and, **10**:61–62
Glow-in-the-dark items, green color
    of, **9**:139–141
Glue
  stickiness in the bottle, **7**:18
  "Super" versus conventional,
    6:145–146
  "Super," and Teflon, **7**:128–129
  taste of postage stamp, **2**:182
Glutamates, MSG and, **2**:168–171
Gnats, swarming of, **4**:3–4
Goalies, banging of sticks by, **7**:116–
    117
Goalposts, football, tearing down of,
    7:181; **8**:213–217
"Goat," **11**:21
Goats, kids versus adult, **7**:64–65
"Gobbledygook," **11**:86–87
Gold bullion, shape of, **8**:32–33

Golf
  origins of eighteen holes, **2**:54
  origins of term "fore," **2**:34
  pain of mishits, **4**:118–119
Golf balls
  dimples on, **3**:45–46
  veering toward ocean while putting,
    6:107–108
Golfers and sunglasses, **9**:158–162
Gondolas, black color of, **4**:86–87
"Good Friday," origins of term,
    8:108–109; **10**:265
Goofy
  identify of, **3**:64–65
  marital status of, **7**:102
Goofy, Jr., origins of, **7**:102
Goosebumps, faces and, **2**:8–9
Gorillas, chest pounding of, **8**:53–55
Gowns, and caps, at graduations,
    6:99–102
Grades in school, E, **3**:198; **4**:206–
    209
Grading of cigarettes, **6**:112–113
Grading of eggs, meaning of, **4**:136–
    137
Graduations, military academy, **2**:20–
    21
"Grandfather" clock, origins of term,
    4:178
Grape jellies, color of, **7**:142–143
Grapefruit, sweetness of, canned ver-
    sus fresh, **1**:199
"Grape-nuts," **11**:171
Grapes, raisins and, **2**:218–219
Gravel and placement on flat roofs,
    6:153–154
Graves, depth of, **7**:14–15
Gravy skin loss, when heated, **6**:58
"Gravy train," **11**:62
Grease, color of, **5**:182
Grecian Formula, process of, **8**:110–
    111
Greek names of fraternities and soror-
    ities, **10**:94–98

Green beans, "French" style, **10**:125–126

Green color of glow-in-the-dark items, **9**:139–141

Green lights, versus red lights, on boats and airplanes, **4**:152–153

"Green with envy," **11**:188

"Green" cards, color of, **7**:61–63

"Greenhorn," **11**:189

Greeting cards, shape of, **6**:70–71

Gretzky, Wayne, hockey uniform of, **2**:18; **10**:279

Grimace, identity of McDonald's, **7**:173

Grocery coupons, cash value of, **5**:7–9

Grocery sacks, names on, **2**:166–167

Grocery stores and check approval, **8**:245–246

Groom, carrying bride over threshold by, **4**:159

Growling of stomach, causes of, **4**:120–121

Guitar strings, dangling of, **8**:11–13; **10**:276

Gulls, sea, in parking lots, **6**:198–199; **10**:254–256

Gum, chewing
  water consumption and hardening of, **10**:236–237
  wrappers of, **8**:111–112

"Gunny sacks," **11**:195

"Guy," **11**:151–152

"Habit [riding costume]," **11**:123

"Hackles," **11**:6–7

Hail, measurement of, **5**:203; **6**:239–240; **7**:234–235; **8**:236–237

Hair
  blue, and older women, **2**:117–118
  growth of, after death, **4**:163–164
  length of, in older women, **7**:179; **8**:192–197
  mole, color of, **8**:167–169
  parting, left versus right, **1**:116

Hair color, darkening of, in babies, **10**:209

Hair spray, unscented, smell of, **2**:184

Hairbrushes, length of handles on, **7**:38–39

Hairs in mouth, gagging on, **7**:76–77

Hairy ears in older men, **2**:239; **3**:231–233; **5**:227–228

Half dollars, vending machines and, **3**:54–56

"Halfback," **11**:138

Half-mast, flags at, **10**:36–38

Half-moon versus quarter moon, **7**:72–73

Half-numbers in street addresses, **8**:253

Halibut, coloring of, **3**:95–96

Halloween, Jack-o'-lanterns and, **4**:180–181

Halogen light bulbs, touching of, **5**:164

Ham
  checkerboard pattern atop, **7**:66–67
  color of, when cooked, **7**:15–16
  Easter and consumption of, **1**:151–152

"Ham [actor]," **11**:170–171

Hamburger buns, bottoms of, **2**:32–34

"Hamburger," origins of term, **4**:125

"Hamfatter," **11**:170–171

Hand dryers in bathrooms, **8**:174–176; **10**:266–267

Hand positions in old photographs, **7**:24–26

Handles versus knobs, on doors, **7**:148–149

Handwriting, teaching of cursive versus printing, **7**:34–37

"Hansom cab," **11**:63

Happy endings, crying and, **1**:79–80

Hard hats
  backward positioning of, in iron-workers, **4**:94
  exterminators and, **2**:51

Hard-boiled eggs, discoloration of, **3**:34
Hat tricks, in hockey, **2**:165–166
Hats
  cowboy, dents on, **7**:249–250
  declining popularity, **5**:202; **6**:227–
    231; **7**:233; **7**:249
  dents in cowboy, **5**:6; **6**:274
  holes in sides of, **5**:126
  numbering system for sizes, **4**:110
Haystacks, shape of, **6**:47–48; **8**:265–
  266
"Hazard [dice game]," **11**:134
"Head [bathroom]," **11**:48
"Head honcho," **11**:39
Head injuries, "seeing stars" and,
  10:156–158
Head lice, kids and, **10**:225–227
Headaches and aspirin, **7**:100–102
Headbands on books, **7**:126–127
Headlamps, shutoff of automobile,
  7:92–93
"Heart on his sleeve," **11**:128
Hearts, shape of, idealized versus real,
  4:199; **5**:220–221; **6**:260; **7**:229–
  230; **8**:234; **9**:234
Heat and effect on sleep, **6**:137–138
"Hector," **11**:155
"Heebie jeebies," **11**:40
Height
  clearance signs on highways, **8**:156–
    158
  of elderly, **6**:250
  restrictions on fences, **2**:28–30
  voice pitch and, **2**:70
Heinz ketchup labels, **8**:150–151
Helicopters, noise of, **8**:164–166
Helium and effect on voice, **5**:108–
  109
Hellman's Mayonnaise, versus Best
  Foods, **1**:211–214
"Hem and haw," **11**:195–196
"Hep," **11**:52–53
Hermit crabs, bathroom habits of,
  7:74–75

Hernia exams and "Turn your head
  and cough," **5**:114–115
"Heroin," **11**:77
"High bias," versus "low bias," on au-
  dio tape, **4**:153–154
"High jinks," **11**:41
High-altitude tennis balls, **8**:80
"Highball," **11**:175–176
Highways
  clumping of traffic, **4**:165–167;
    7:247
  curves on, **7**:121–122
  interstate, numbering system, **4**:66–
    67
  traffic jams, clearing of, **1**:25–26
  weigh stations, **4**:193–194
"Hillbilly," **11**:148
Hills, versus mountains, **3**:97–98;
  8:252
"Hip," **11**:52–53
"Hobnob," **11**:40
"Hobson's choice," **11**:155–156
Hockey
  banging of sticks by goalies, **7**:116–
    117
  hat trick, **2**:165–166
  Montreal Canadiens uniforms,
    5:165; **7**:242
  Wayne Gretzky's uniform, **2**:18;
    10:279
"Hold a candle," **11**:71
"Holding the bag," **11**:75
Holes
  in barrels of cheap pens, **4**:111
  in elevator doors, **8**:170–171
  in fly swatters, **4**:31–32
  in ice cream sandwiches, **8**:128
  in needles and syringes, **10**:57–59
  in pasta, **4**:28
  in saltines, **8**:129
  in thimbles, **10**:63–64
  in wing-tip shoes, **8**:44
  on bottom of soda bottles, **6**:187–
    188

recycling of, in loose-leaf paper,
7:105–106
refilling of dirt, 7:48–49
"Holland," versus "Netherlands,"
2:65–66
Home plate, shape of, in baseball,
5:131
"Honcho," 11:39
Honey, spoilage of, 4:177–178
Honey roasted peanuts, banning of,
on airlines, 4:13–14
Honking in geese during migration,
7:108
"Honky," 11:77
"Hoodwink," 11:121
"Hoosiers," 11:148–149
"Horsefeathers," 11:40
Horses
measurement of heights of, 5:60–
61
posture in open fields, 3:104; 5:241
shoes, 3:156
sleeping posture, 2:212
vomiting, 6:111–112; 7:248
Hospital gowns, back ties on, 5:132–
134
Hospitals and guidelines for medical
conditions, 4:76–78
Hot dog buns
number of, in package, 2:232–235
slicing of, 5:161
Hot dogs, skins of, 5:54
Hot water, noise in pipes of, 2:199–
200
Hotels
amenities, spread of, 6:118–121
number of towels in rooms, 4:56–
57
plastic circles on walls of, 3:117
toilet paper folding in bathrooms
of, 3:4
"Hotsy totsy," 11:40
Houses, settling in, 6:32–34
"Hue and cry," 11:112

"Humble pie," 11:169
Humidity, relative, during rain,
1:225–226
Humming, power lines and, 10:259
Hurricane, trees and, 3:68–69
"Hurricanes" as University of Miami
nickname, 8:171–172
"Hut," origins of football term, 6:40;
9:294–295
Hydrants, fire, freezing water in,
10:11
Hypnotists, stage, techniques of,
1:180–191

"I [capitalization of]," 11:55
"I could care less," 11:78
"I" before "e," in spelling, 6:219;
7:209; 8:240–245
Ice
fizziness of soda, 9:24–25
formation on top of lakes and
ponds, 5:82–83
holes and dimples in, 9:147–148
in urinals, 10:232–234
Ice cream
black specks in, 8:132–133
cost of cones versus cups, 1:133–
135
pistachio, color of, 7:12–13
thirstiness, 5:202; 6:236–237
Ice cream and soda, fizziness of, 9:27
Ice cream sandwiches, holes in, 8:128
Ice cubes
cloudy versus clear, 3:106–107;
5:242
shape of, in home freezers, 5:103–
104
Ice rinks, temperature of resurfacing
water in, 10:196–198
Ice skating, awful music in, 1:102–105
Ice trays in freezers, location of,
10:92–93
Icy roads, use of sand and salt on,
2:12–13

Ignitions
  automobile, and headlamp shutoff,
    7:92–93
  key release button on, **5**:169
Imperial gallon, versus American gal-
    lon, **6**:16–17
"In like Flynn," **11**:157
"In the nick of time," **11**:158
Index fingers and "Tsk-Tsk," stroking
    of, **4**:198; **5**:209–210
"Indian corn," **11**:146
"Indian pudding," **11**:146
"Indian summer," **11**:146
Indianapolis 500, milk consumption
    by victors in, **8**:130–131
"Inflammable," versus "flammable,"
    2:207–208
Ink
  color of, in ditto masters, **6**:133–134
  newspaper, and recycling, **7**:139–
    140
Insects
  attraction to ultraviolet, **8**:158–159
  aversion to yellow, **8**:158–159
  flight patterns of, **7**:163–164
  in flour and fruit, **4**:89–90
Insects [see specific types]
Insufficient postage, USPS procedures
    for, **4**:149–151
Interstate highways, numbering sys-
    tem of, **4**:66–67
Introductions in books, versus fore-
    words and prefaces, **1**:72–73
Irish names, "O' " in, **8**:135–136
Irons, permanent press settings on,
    3:186–187
Ironworkers, backwards hard hat
    wearing of, **4**:94
Irregular sheets, proliferation of,
    1:145–147
IRS and due date of taxes, **5**:26–29
IRS tax forms
  disposal of, **8**:143–144
  numbering scheme of, **4**:9–10

"Italian" bread, versus "French,"
    7:165–166; **8**:261–262
Itching, reasons for, **1**:172–173
Ivory soap, purity of, **2**:46–47

"J" Street, Washington D.C. and, **2**:71
"Jack [playing card]," **11**:135
Jack Daniel's and "Old No. 7," **8**:144–
    145
"Jack," "John" versus, **2**:43
Jack-o'-lanterns, Halloween and,
    4:180–181
Jams, contents of, **6**:140–141
Japanese
  baseball uniforms, **10**:207–208
  boxes, yellow color of, **7**:130–131
  flags, red beams and, **10**:151–155
Jars, food, refrigeration of opened,
    6:171–172
"Jaywalking," **11**:22–23
Jeans
  blue, orange thread and, **9**:74
  Levis, colored tabs on, **6**:59–61
  origin of "501" name, **6**:61
  sand in pockets of new, **7**:152
"Jeans [pants]," **11**:124
"Jeep," **11**:61
Jellies, contents of, **6**:140–141
Jellies, grape, color of, **7**:142–143
Jello-O, fruit in, **3**:149–150
Jeopardy, difficulty of daily doubles
    in, **1**:33–35
"Jerkers," **11**:176
Jet lag, birds and, **3**:33–34
"Jetsam," versus "flotsam," **2**:60–61
"Jig is up," **11**:7
Jigsaw puzzles, fitting pieces of, **9**:3–4
Jimmies, origins of, **10**:165–168
"Jink," **11**:41
"John," versus "Jack," **2**:43
Johnson, Andrew, and 1864 election,
    8:85–87
"Joshing," **11**:158–159
Judges and black robes, **6**:190–192

MASTER INDEX OF IMPONDERABILITY

Measuring spoons, inaccuracy of, 1:106–107

Meat
children's doneness preferences, 5:230–231; 6:252–253; 9:273–274
national branding, 1:227–231; 9:287
red color of, 8:160–161

Meat loaf, taste in institutions, 5:203; 6:243; 7:235–236

Medals, location of on military uniforms, 2:223–224

Medical conditions, in hospitals, guidelines for, 4:75–76

Medicine bottles, cotton in, 3:89–90

Memorial Day, Civil War and, 3:168–169

Men
dancing ability of, 6:218; 7:199–202; 8:239–240
feelings of coldness, 6:218; 7:198–199
remote controls and, 6:217; 7:193–196

Menstruation, synchronization of, in women, 4:100–102

Menthol, coolness of, 5:192

Meter, origins of, 2:200–202

Miami, University of
football helmets, 8:171–172
"Hurricanes" nickname, 8:171–172

Mickey Mouse, four fingers of, 3:32; 6:271

Microphones, press conferences and, 2:11–12

Migration of birds, 9:91–94

Mile, length of, origins of, 1:241–242

Military salutes, origins of, 3:147–149

Milk
as sleep-inducer, 7:17
fat content in lowfat, 7:60–61
in refrigerators, coldness of, 5:4–5
Indianapolis 500, 8:130–131

national brands, 1:227–231; 9:287
plastic milk containers, 7:61; 10:262–263
single serving cartons of, 7:137–138
skim versus nonfat, 7:59
skin on, when heated, 6:58

Milk cartons
design of, 5:112; 9:289–290
difficulty in opening and closing of, 1:243–246; 5:243

Milk cases, warnings on, 5:43–44

Milk Duds, shape of, 8:81–82

Millimeters, as measurement unit for film, 1:44

"Mind your P's and Q's," 11:88–89

Mint flavoring on toothpicks, 4:153

Mint, U.S., and shipment of coin sets, 5:32

Minting of new coins, timing of, 3:128

Mirrors in bars, 10:14–17

Mirrors, rear-view, 4:185–186

Mistletoe, kissing under, origins of, 4:106–107

Mobile homes, tires atop, in trailer parks, 6:163–164

Mole hair, color of, 8:167–169

Money, U.S.
color of, 3:83–84
stars on, 3:180–182

Monitors, computer, shape of, 6:129–131

Monkeys, hair picking of, 3:26–27

Monopoly, playing tokens in, 10:21–23

Montreal Canadiens, uniforms of, 5:165; 7:242

Moon
apparent size of, at horizon, 2:202–204
effect on lakes and ponds, 5:138–139
official name, 5:19–20
quarter-, vs. half-, 7:72–73

"No bones about it," **11**:49

"No Outlet" signs, versus "Dead End" signs, **4**:93

Noise, traffic, U.S. versus foreign countries, **4**:198

North Carolina, University of, and Tar Heels, **8**:76–77

North Pole
directions at, **10**:243
telling time at, **10**:241–243

Nose rings and bulls, **10**:147–148

Noses
clogged nostrils and, **3**:20–21
runny, in cold weather, **10**:146–147
runny, kids versus adults, **9**:89–90
wet, in dogs, **4**:70–73

Nostrils, clogged, **3**:20–21

Notches on bottom of shampoo bottles, **10**:29–30

Notre Dame fighting Irish, **10**:115–117

NPR radio stations, low frequency numbers of, **10**:181–183

Numbers, Arabic, origins of, **3**:16–17

Nutrition labels, statement of fats on, **6**:142–143

Nuts
Brazil, in assortments, **7**:145–147
Macadamia shells, **8**:262
peanuts in plain M&M's, **7**:239
peanuts, and growth in pairs, **7**:34

"O'" in Irish names, **8**:135–136

Oakland A's, elephant on uniforms of, **6**:14–15

Oboes, use of as pitch providers, in orchestras, **4**:26–27

Occupancy, maximum, in public rooms, **10**:158–160

Oceans
boundaries between, **10**:74–76
color of, **2**:213
salt in, **5**:149–150
versus seas, **5**:30–32

Octopus throwing, Detroit Red Wings and, **9**:183–186

"Off the schneider," **11**:136

Oh Henry, origins of name of, **8**:83–84

Oil
automotive, after oil change, **5**:184–185; **7**:240–241
automotive, grades of, **3**:182–183

"Okay," thumbs-up gesture as, **1**:209–210

Oktoberfest, September celebration of, **9**:156–157

Old men
hairy ears and, **2**:239; **3**:231–233; **5**:227–228
pants height and, **2**:171–172; **6**:274

"Old No. 7" and Jack Daniel's, **8**:144–145

"Old Zealand," versus New Zealand, **4**:21–22

Olive Oil, virgin, **3**:174–175

Olives, green and black, containers of, **1**:123–127

"On pointe" and ballet, **8**:69–72

"On tenterhooks," **11**:10–11

"On the Q.T.," **11**:59

"Once in a blue moon," **11**:12

"One fell swoop," **11**:197

One Hour Martinizing, **3**:28–29

One-hour photo processing, length of black-and-white film and, **4**:39

Onions and crying, **9**:169–170

Orange coffee pots, in restaurants, **6**:67–69

Orange juice
price of fresh versus frozen, **5**:155–156
taste of, with toothpaste, **10**:244–246

Orange thread in blue jeans, **9**:74

Oranges, extra wedges of, **4**:175–176

Oranges, mandarin, peeling of, **8**:106–107

Penguins
  frostbite on feet, **1**:217–218
  knees, **5**:160
Penicillin and diet, **8**:95–96
Penmanship of doctors, bad, **5**:201;
    6:221–225; **7**:232–233; **8**:235
Pennies
  lettering on, **7**:5
  smooth edges of, **1**:40–41
  vending machines and, **3**:54–56
Pennsylvania Dept. of Agriculture,
    registration, baked goods, **2**:121–
    122
Penny loafers, origins of, **8**:43–44
Pens
  disappearance of, **4**:199; **5**:222–
    223; **6**:260–261; **7**:230–231;
    8:234
  holes in barrel of cheap, **4**:111
  ink leakages in, **4**:112–113
Pepper
  and salt, as condiments, **5**:201;
    6:225–226; **8**:235–236
  and sneezing, **8**:61
  white, source of, **2**:135–136
Pepsi-Cola, trademark location of,
    5:115–116
Perfume
  color of, **9**:19
  wrists and, **6**:90
Periods in telegrams, **3**:77–78
Permanent press settings on irons,
    3:186–187
Permanents, pregnancy and, **3**:170–
    171
Perpetual care, cemeteries and,
    2:221–222
"Peter out," **11**:163–164
Phantom limb sensations, amputees
    and, **1**:73–75
Pharmacists and raised platforms,
    8:5–7
Philips screwdriver, origins of, **2**:206–
    207

Philtrums, purpose of, **6**:43; **8**:266–
    267
Photo processing, one-hour, length of
    black-and-white film and, **4**:39
Photographs
  poses in, **6**:218
  red eyes in, **4**:68–69
  stars in space and, **10**:213–215
Photography
  color of cameras, **7**:246
  hand position of men in old, **7**:24–
    26
  hands on chins in, **7**:203–207
  Polaroid prints, flapping of, **7**:175–
    176
  smiling in old photographs, **7**:19–
    23
Physical exams, back tapping during,
    3:145–146
"Pi" as geometrical term, **5**:80–81
Piano keys, number of, **10**:7–9
"Pig in a poke," **11**:25
Pigeons
  baby, elusiveness of, **1**:254; **10**:253–
    254
  loss of toes, **7**:166
  whistling sound in flight, **7**:58
Pigs
  curly tails of, **4**:199; **5**:218–219;
    7:228–229
  pink hair color of, **8**:98
  roasted, and apples in mouths of,
    7:159; **10**:274–275
Pilgrims, buckled hats of, **10**:47–51
Pillow tags, label warnings of, **2**:1–2
Pilots
  and dimming of interior lights,
    8:24–25
  female, dearth of, on airlines,
    1:131–133
"Pin money," **11**:115
Pine nuts, shelling of, **2**:94
Pine trees, construction sites and,
    2:147–148

Pineapple in gelatin, **3**:149–150
Pinholes, on bottle caps, **2**:223
Pink as color for baby girls, **1**:29
"Pink lady," **11**:190–191
Pink stripes on magazine labels, **8**:96–97
"Pinkie," **11**:190–191
Pins in men's dress shirts, **4**:29–30
"Pipe down," **11**:13
Pipes, kitchen, shape of, **4**:82–83
Pirates
  earrings on, **9**:43–45; **10**:272–273
  walking the plank, **9**:37–42; **10**:273–274
Pistachio ice cream, color of, **7**:12–13
Pistachios, red color of, **1**:26–28
Pita bread, pockets in, **6**:98
Pitcher's mounds
  location of, **5**:181
  rebuilding of, **9**:195–198
Pittsburgh Steelers, emblems on helmets of, **7**:67–68
Planets, twinkling of, at night, **4**:50–51
Plastic bottles, beer and, **7**:161–162
Plastic cups, shape of, **9**:289
Plastic deer ornaments on lawns, **9**:262–264
Plastics, recyclable, numbers on, **6**:155–156
Plates
  repositioning of, **8**:184; **9**:238–242
  round shape of dinner, **8**:162–164
Plots, farm, circular shape of, **7**:118–119
Plug prongs
  holes at end of, **5**:94–95
  three prongs versus two prongs, **5**:191
Plum pudding, plums in, **5**:49
Plumbing
  kitchen, shape of, **4**:82–83
  sound of running water, **3**:239–240
Pockets in pita bread, **6**:98

Poison ivy, grazing animals and, **3**:86–87
Polaroid prints, flapping of, **7**:175–176
Poles
  directions at, **10**:243
  telling time at North and South, **10**:241–243
Pole-vaulting
  preparation for different heights, **9**:97–101
  women and, **9**:102–107
Police
  and crowd estimates, **1**:250–253
  flashlight grips, **10**:30–32
  radar and speed measurement, **8**:88–91
Police car beacons, colors on, **7**:135–137
Police dogs, urination and defecation of, **3**:67–68
Policemen and mustaches, **6**:219; **7**:218–220; **8**:246–247; **9**:278
Ponds
  effect of moons on, **5**:138–139
  fish returning to dried, **3**:15–16; **10**:256
  ice formations on, **5**:82–83
  versus lakes, **5**:29–30; **7**:241
  versus lakes, level of, **9**:85–86
Poodles, wild, **6**:207–209
Pool balls, dots on, **10**:237–240
"Pop goes the weasel," **11**:14
Popcorn
  "gourmet" versus regular, **1**:176
  popping in-house, in movie theaters, **1**:45–50
  versus other corns, **3**:142–143
Popes
  name change of, **10**:17–18
  white skullcap of, **10**:80–81
  white vestments of, **10**:79–80
Popping noise of wood, in fires, **4**:10
Pork and beans, pork in, **2**:19

"Port," **11**:65–66

"Porthole," **11**:65–66

Post office, translation of foreign mail and, **3**:133

Postage and ripped stamps, **8**:62

Postage Stamps
leftover perforations of, **4**:179
taste of, **2**:182

Postal Service, U.S., undeliverable mail and, **5**:13–14

Pot pies, vent holes in, **6**:28

Potato chips
bags, impossibility of opening and closing, **9**:117–118
curvy shape, **9**:115–116
green tinges on, **5**:136–137; **6**:275
price of, versus tortilla chips, **5**:137–138

Potato skins, restaurants and, **3**:12–13

Potatoes, baked, and steak houses, **6**:127–129

Potholes, causes of, **2**:27

"Potter's field," **11**:198

Power lines
humming of, **9**:165–168; **10**:259
orange balls on, **4**:18–19

Prefaces in books, versus introductions and forewords, **1**:72–73

Pregnancy, permanents and, **3**:170–171

Pregnant women, food cravings of, **10**:183–185.

Preserves, contents of, **6**:140–141

Press conferences, microphones in, **2**:11–12

"Pretty kettle of fish," **11**:178

"Pretty picnic," **11**:178

Pretzels, shape of, **6**:91–92

Priests, black vestments and, **10**:77–79

Priority mail, first class versus, **3**:166–167

Prisoners and license plate manufacturing, **8**:137–139

Prohibition, liquor production of distilleries during, **9**:54–56

Pronunciation, dictionaries and, **10**:169–179

"P's and Q's," **11**:88–89

Pubic hair
curliness of, **5**:177–178
purpose of, **2**:146; **3**:242–243; **6**:275–276

Public buildings, temperatures in, **8**:184

Public radio, low frequency numbers of, **10**:181–183

Pudding, film on, **6**:57

Punts, measurement of, in football, **3**:124–125

Purple
Christmas tree lights, **6**:185–186; **9**:293; **10**:280
paganism, **9**:292–93
royalty and, **6**:45–46

"Put up your dukes," **11**:137

Putting, veering of ball toward ocean when, **6**:107–108

"Q.T.," **11**:59

"Qantas," spelling of, **8**:134–135

Q-Tips, origins of name, **6**:210–211

"Quack [doctor]," **11**:45

"Quarterback," **11**:138

Quarterbacks and exclamation, "hut," **6**:210

Quarter-moons versus half moons, **7**:72–73

Quarts and gallons, American versus British, **4**:114–115

Queen-size sheets, size of, **3**:87–88

Rabbit tests, death of rabbits in, **7**:69–71

Rabbits and nose wiggling, **5**:173–174

Racewalking, judging of, **9**:20–23

Racquetballs, color of, **4**:8–9

Radar and police speed detection, 8:88–91

Radiators and placement below windows, 9:128–130

Radio
beeps before network news, 1:166–167
FM, odd frequency numbers of, 10:59–60
lack of song identification, 7:51–57
public, low frequency numbers, 10:181–183

Radio Shack and lack of cash registers, 5:165–166

Radios
battery drainage, 10:259–260
lingering sound of recently unplugged, 4:47

Railroad crossings and "EXEMPT" signs, 5:118–119

Railroads, width of standard gauges of, 3:157–159

Rain
butterflies and, 4:63–64
fish biting in, 10:131–138
measurement container for, 4:161–163
smell of impending, 6:170–171; 7:241

Raincoats, dry-cleaning of, 2:216–217

"Raining cats and dogs," 11:26

"Raise hackles," 11:6–7

Raisins
cereal boxes and, 2:123
seeded grapes and, 2:218–219

Ranchers' boots on fence posts, 2:77–81; 3:243–245; 4:231; 5:247–248; 6:268; 7:250; 8:253; 9:298–299; 10:251–253

Razor blades, hotel slots for, 2:113

Razors, men's versus women's, 6:122–123

"Read the riot act," 11:89

"Real McCoys," 11:164–165

Rear admiral, origins of term, 5:25

Rear-view mirrors, day/night switch on automobile, 4:185–186

Receipts, cash register, color of, 4:143

Records, vinyl, speeds of, 1:58–61

Recreational vehicles and wheel covers, 7:153

Recyclable plastics, numbers on, 6:155–156

Recycling of newspaper ink, 7:139–140

Red
color of beef, 8:160–161
eyes in photographs, 4:68–69
paint on coins, 7:117

"Red cent," 11:191

"Red herring," 11:185–186

Red lights, versus green lights, on boats and airplanes, 4:152–153

"Red tape," 11:187

Red Wings, Detroit, octopus throwing and, 9:183–186

"Red-letter day," 11:186–187

Redshirting in college football, 7:46–48

Refrigeration of opened food jars, 6:171–172

Refrigerators
location of freezers in, 2:230–231
smell of new, 8:91–92
thermometers in, 10:85–87

Relative humidity, during rain, 1:225–226

Remote controls, men versus women and, 6:217; 7:193–196

Repair shops, backlogs and, 4:45–47

Restaurants
coffee, versus home-brewed, 7:181
vertical rulers near entrances of, 7:95–96

Restrooms, group visits by females to, 6:217; 7:183–192; 8:237–238; 9:277–278

Revolving doors, appearance of, in big cities, **4**:171–173

Reynolds Wrap, texture of two sides of, **2**:102

Rhode Island, origins of name, **5**:21–22

Ribbons, blue, **3**:57–58

Rice cakes, structural integrity of, **10**:9–11

Rice Krispies
noises of, **3**:165; **5**:244–245
profession of Snap!, **8**:2–4

"Right wing," **11**:116

"Rigmarole," **11**:81–82

Rings, nose, bulls and, **10**:147–148

Rinks, ice, temperature of resurfacing water in, **10**:196–198

"Riot Act [1716]," **11**:89

Roaches
automobiles and, **7**:3–4; **8**:256–257; **9**:298
position of dead, **3**:133–134; **8**:256
reactions to light, **6**:20–21

Roads
blacktop, coloring of, **5**:22–23
fastening of lane reflectors on, **5**:98–99
versus bridges, in freezing characteristics, **5**:193

Robes, black, and judges, **6**:190–192

Rocking in zoo animals, **10**:279

Rodents and water sippers, **6**:187

Roller skating rinks, music in, **2**:107–108; **6**:274–275

Rolls
coldness of airline, **3**:52–53
versus buns, **8**:65–66

Roman chariots, flimsiness of, **10**:105–107

Roman numerals
calculations with, **3**:105–106
copyright notices in movie credits and, **1**:214–216
on clocks, **5**:237–238

Roofs, gravel on, **6**:153–154

Roosevelt, Teddy, and San Juan Hill horses, **4**:49

Roosters, crowing and, **3**:3

Root beer
carbonation in, **1**:87–88
foam of, **8**:93–94

Rubble, Betty
nonappearance in Flintstones vitamins, **6**:4–5; **9**:285–286
vocation of, **7**:173–174

Ruins, layers of, **2**:138–140

Rulers, vertical, in restaurant entrances, **7**:95–96

Run amok, **11**:68

Runny noses
cold weather and, **10**:146–147
kids versus adults, **9**:89–90

Rust, dental fillings and, **10**:41–42

RVs and wheel covers, **7**:153

"Rx," **11**:59

S.O.S Pads, origins of, **8**:103–104

"Sacked [fired]," **11**:75

Sacks, paper
jagged edges on, **6**:117–118
names on, **2**:166–167

Safety caps, aspirin, 100–count bottles of, **4**:62

Safety pins, gold versus silver, **9**:87–88

Saffron, expense of, **1**:129–130

Sailors, bell bottom trousers and, **2**:84–85

"Salad days," **11**:178

Salads, restaurant, celery in, **6**:218; **7**:207–209

Saloon doors in Old West, **3**:198

Salt
and pepper, as table condiments, **5**:201; **6**:225–226; **8**:235–236
in oceans, **5**:149–150
packaged, sugar as ingredient in, **4**:99

round containers and, **10**:149–150
storage bins on highway and,
10:216–218
versus sand, to treat icy roads,
2:12–13
Salutes, military, origins of, **3**:147–149
San Francisco, sourdough bread in,
2:180–181
Sand
in pockets of new jeans, **7**:152
storage bins on highway and,
10:216–218
versus salt, to treat icy roads, **2**:12–
13
Sandbags, disposal of, **10**:193–195
Sardines, fresh, nonexistence in super-
markets, **1**:76–78
"Sawbuck [ten-dollar bill]," **11**:119–
120
Sawdust on floor of bars, **10**:118–120
Scabs, itchiness of, **5**:125–126
Scars, hair growth and, **2**:186
"Schneider," **11**:136
School clocks, backward clicking of
minute hands in, **1**:178–179
Schools, CPR training in, **7**:31–33
Scissors, sewing, and paper cutting,
8:131–132
"Scot-free," **11**:143
"Scotland Yard," **11**:144
Scotsmen and kilts, **7**:109–110
Screen doors, location of handles on,
7:91
Screwdrivers, reasons for Philips,
2:206–207
Scuba masks, spitting into, **9**:30–34
Sea gulls in parking lots, **6**:198–199;
10:254–256
Sea level, measurement of, **4**:154–155
Seas versus oceans, differences be-
tween, **5**:30–32
Seat belts
and shoulder straps in airplanes,
8:141–142

in buses, **1**:84–85
in taxicabs, **1**:85
Secretary as U.S. government depart-
ment head designation, **3**:121–
122
"Seed [tournament ranking]," **11**:141–
142
"Seeing stars," head injuries and,
10:156–158
Self-service versus full-service, pricing
of, at gas stations, **1**:203–209
"Semi," origins of term, **2**:179
"Semimonthly," **11**:194
"Semiweekly," **11**:194
Serrated knives, lack of, in place settings, **4**:109–110
Settling in houses, **6**:32–34
Seven-layer cakes and missing layers,
6:80–81
Seventy-two degrees, human comfort
at, **2**:178–179
Sewing scissors and paper cutting,
8:131–132
Shampoo bottles, notches on bottom
of, **10**:29–30
Shampoo labels, "FD&C" on label of,
4:163
Shampoos
colored, white suds and, **4**:132–133
lathering of, **5**:44–45
number of applications, **1**:90–93
Shaving of armpits, women and,
2:239; **3**:226–229; **6**:249
Sheets
irregular, proliferation of, **1**:145–147
queen-size, size of, **3**:87–88
Sheriffs' badges, shape of, **5**:73–74
Shirts
buttons on men's versus women's,
2:237–238; **3**:207–209; **5**:226;
7:223
men's, pins in, **4**:29–30
single-needle stitching in, **6**:51
starch on, **3**:118

Shoe laces
  length in athletic shoes, **8**:41–42
  untied, in shoe stores, **8**:40
Shoe sizes, differences between,
  1:65–70
Shoes
  lace length in shoe stores, **8**:41–42
  layers on, **5**:59
  of deceased, at funerals, **7**:153–154
  penny loafers, **8**:43–44
  single, on side of road, **2**:236–237;
    3:201–207; **4**:232–233; **5**:225–
    226; **6**:245–248; **7**:221–222;
    8:228–230; **9**:271–272
  tied to autos, at weddings, **1**:235–
    238
  uncomfortable, and women, **1**:62–
    64
  untied laces in stores, **8**:40
  wing-tip, holes in, **8**:44
"Shoofly pie," **11**:179
Shopping, female proclivity toward,
  7:180; **8**:205–209
Shopping malls, doors at entrance of,
  6:180–181
"Short shrift," **11**:15
Shoulder straps and seat belts in air-
    planes, **8**:141–142
Shredded Wheat packages, Niagara
    Falls on, **5**:100–101
"Shrift," **11**:15
Shrimp, baby, peeling and cleaning of,
  5:127
"Shrive," **11**:15
"Siamese twins," **11**:151
"Sic," as dog command, **5**:51
Side vents in automobile windows,
  6:13–14
"Sideburns," **11**:126–127
Sidewalks
  cracks on, **3**:176–178
  glitter on, **7**:160; **10**:61–62
Silica gel packs in electronics boxes,
  6:201

Silos, round shape of, **3**:73–74; **5**:245;
    10:260–261
Silver fillings, rusting of, **10**:41–42
Silverstone, versus Teflon, **2**:3
Singers, American accents of foreign,
  4:125–126
Single-needle stitching in shirts, **6**:51
Sinks, overflow mechanisms on,
  2:214–215
Siskel, Gene, versus Roger Ebert,
    billing of, **1**:137–139
Skating music, roller rinks and,
  2:107–108; **6**:274–275
Skating, figure, and dizziness, **5**:33–35
Ski poles, downhill, **3**:69
"Skidoo," **11**:100
Skunks, smell of, **10**:88–91
Skyscrapers, bricks in, **6**:102–103
Skytyping, versus skywriting, **9**:17–18
Skywriting
  techniques of, **9**:12–16
  versus skytyping, **9**:17–18
Sleep
  babies and, **6**:56–57
  drowsiness after meals, **6**:138–139
  eye position, **6**:146
  heat and effect on, **6**:137–138
  twitching during, **2**:67
"Slippery When Wet" signs, location
    of, **7**:167–168
"Small fry," **11**:179
Smell of impending rain, **6**:170–171;
  7:241
Smiling in old photographs, **7**:19–23
Smoke from soda bottles, **8**:148
Snack foods and prepricing, **3**:79–
  80
Snake emblems on ambulances,
  6:144–145; **7**:239–240
Snakes
  sneezing, **7**:98
  tongues, **2**:106; **10**:278
Snap! [of Rice Krispies], profession of,
  8:2–4

"Snap! Crackle! And Pop!" of Rice
   Krispies, **3**:165
Sneezing
   eye closure during, **3**:84–85
   looking up while, **2**:238
   pepper and, **8**:61
   snakes and, **7**:98
Snickers, wavy marks on bottom of,
   6:29–31
Sniffing, boxers and, **2**:22–23;
   10:256–258
Snoring, age differences and, **10**:126–
   127
Snow and cold weather, **3**:38
"Snow" on television, **9**:199–200
Soap, Ivory, purity of, **2**:46
Soaping of retail windows, **9**:265–267
Soaps, colored, white suds and,
   4:132–133
Social Security cards, lamination of,
   5:140–141
Social Security numbers
   fifth digit of, **8**:100–102
   reassignment of, **5**:61–62
   sequence of, **3**:91–92; **6**:267
Socks
   angle of, **3**:114–115
   disappearance of, **4**:127–128;
   5:245–246; **6**:272; **8**:266
   men's, coloring of toes on, **4**:19–
   20
"Soda jerk," **11**:176
Soft drinks
   bottles, holes on bottom of, **6**:187–
   188
   brominated vegetable oil in, **6**:96–
   97
   calorie constituents, **6**:94–95
   effect of container sizes and taste,
   2:157–159
   filling of bottles of, **4**:53
   finger as fizziness reduction agent,
   9:28–29
   fizziness in plastic cups, **9**:26

fizziness of soda with ice cream,
   9:27
fizziness over ice, **9**:24–25
freezing of, in machines, **9**:10–11
Kool-Aid and metal containers, **8**:51
machines, "Use Correct Change"
   light on, **9**:186–188
phenylalanine as ingredient in,
   6:96–97
pinholes on bottle caps of, **2**:223
root beer, foam of, **8**:93–94
smoke of, **8**:148
"sodium-free" labels, **4**:87–88
Soles, sunburn and, **8**:63–64
"Son of a gun," **11**:82
Sonic booms, frequency of, **4**:23
Sororities, Greek names of, **10**:94–98
Souffles and reaction to loud noises,
   7:41–42
Soup, alphabet, foreign countries and,
   10:73
Soups and shelving in supermarkets,
   6:26–27
Sour cream, expiration date on, **3**:132
Sourdough bread, San Francisco, taste
   of, **2**:180–181
South Florida, University of, location
   of, **4**:7–8
South Pole
   directions at, **10**:243
   telling time at, **10**:241–243
Sparkling Wine, name of, versus
   champagne, **1**:232–234
Speech, elderly versus younger and,
   6:24–25
Speed limit, 55 mph, reasons for,
   2:143
Speeding and radar, **8**:88–91
Speedometers, markings of, in auto-
   mobiles, **2**:144–145
Spelling, "i" before "e" in, **6**:219;
   7:209; **8**:240–245
Sperm whales, head oil of, **6**:87–89
"Spic and span," **11**:41

Sunburn
   delayed reaction in, **7**:114–116
   palms and soles, **8**:63–64
"Sundae," **11**:181
Sundials, mottoes on, **6**:54–56
Sunglasses and professional golfers,
   9:158–162
Sunrises, timing of, **5**:176–177
Sunsets, timing of, **5**:176–177
"Super" glue and Teflon, **7**:128–129
"Super" glues versus ordinary glues,
   6:145–146
Supermarkets
   check approval policies, **6**:219;
      7:210–217
   public bathrooms in, **6**:157
   shelving of soup, **6**:26–27
Surgeons' uniforms, color of, **2**:86;
   6:269
"Swan song," **11**:27–28
Swarming of gnats, **4**:3–4
Sweating, swimmers and, **10**:261–262
"Sweating" and drinking glasses,
   9:124–125; **10**:261
Swimming
   cats and, **1**:86
   sweating and, **10**:261–262
Swiping of credit cards, **10**:138–141
Swiss cheese
   holes in, **1**:192
   slice sizes of, **9**:142–146
Switches, light, **4**:183–184
Syringes, hole in needle of, **10**:57–59

Tails, curly, pigs and, **4**:199
"Talking a blue streak," **11**:90
Tall old people, rareness of, **3**:229–
   231; **5**:226; **6**:250
Tamers, animal, and kitchen chairs,
   7:9–11
Tape counters, audio and VCR,
   6:272–273
Tar Heels and University of North
   Carolina, **8**:76–77

Taste, sense of, in children versus
   adults, **3**:199; **6**:252–253
Tattoos, color of, **3**:157
"Taw," **11**:9–10
Tax forms
   disposal of, **8**:143–144
   numbering scheme of, **4**:9–10
Taxes, April 15 due date of, **5**:26–29
Taxicabs
   rear windows of, **5**:143–144
   seat belts in, **1**:85
Teachers, apples for, **2**:238; **3**:218–
   220
Teddy bears, frowns of, **10**:160–164
Teeth, silver fillings and, **10**:41–42
Teeth direction of keys, **8**:59–60
"Teetotaler," **11**:42–43
Teflon, stickiness of, **2**:3
Telegrams
   exclamation marks and, **3**:76–77
   periods and, **3**:77–78
Telephone cords, twisting of, **3**:45
Telephone rings, mechanics of,
   4:189–190
Telephones
   911 as emergency number, **5**:145–
      146
   area code numbers, **5**:68–69; **9**:287
   dialing 9 to get outside line, **3**:75–
      76
   fast versus slow busy signals,
      4:182
   holes in mouthpiece, **3**:14–15
   pay, clicking noise in, **4**:97–98
   "Q" and "Z," absence from buttons,
      5:66–67
   rings, mechanics of, **4**:189–190
   third-party conversations, **9**:152–
      155
   three-tone signals, **4**:129–130
   touch tone keypad for, **2**:14–15
   unlisted phone numbers, **9**:45–47
   windowless central offices, **9**:176–
      181

Trains
   backwards locomotives in, **5**:15–16
   "EXEMPT" signs at railroad cross-
      ings, **5**:118–119
Tread, tire, disappearance of, **2**:72–74
Treasurer, U.S., gender of, **8**:45–46
Treasury, printing of new bills by,
   3:126–128
Treble clefs, dots on, **10**:210–213
Trees
   bark, color of, **6**:78–79
   growth in cities, **8**:78–80
   growth on slopes of, **6**:141–142;
      7:244
Triton, orbit pattern of, **4**:117–118
Tropical fish, oxygen in, **7**:84–85
Trucks
   idling of engines of, **7**:150–151
   license plates on, **3**:96; **10**:270–271
   origins of term "semi" and, **2**:179
"Tsk-Tsk," stroking of index fingers
      and, **4**:198
Tuba bells, orientation of, **8**:147–148
Tuesday
   release of CDs, **10**:108–112
   U.S. elections and, **1**:52–54
Tumbleweed, tumbling of, **6**:5–7
Tuna, cat food, cans of, **8**:8–10
Tunnels, ceramic tiles in, **6**:135–136;
      8:257–258
Tupperware and home parties, **3**:25–
      26
"Turkey," in bowling, origin of term,
      4:48–49; **6**:269
Turkeys
   beards on, **3**:99
   white versus dark meat, **3**:53–54
Turn signals in automobiles, clicking
      sounds of, **6**:203
TV Guide, order of listings in, **4**:91–92
Twenty-four second clock in NBA bas-
      ketball, **1**:29–31
Twenty-one as age of majority, **7**:50–
      51

Twenty-one gun salute, origins of,
      2:68–70
"Twenty-three skidoo," **11**:100
Twins, identical, DNA and, **10**:18–20
Twitches during sleep, **2**:67
TWIX cookie bars, holes in, **6**:28–29
"Two bits," origins of term, **2**:191–192
Two by fours, measurement of, **2**:87–
      88
Two-minute warning, football and,
      10:150–151
Typewriter keys, location of, **1**:127–
      128

"U" in University of Miami football
      helmets, **8**:171–172
Ultraviolet and attraction of insects,
      8:158–159
Umpires and home plate sweeping,
      8:27–31
"Uncouth," **11**:13
Underarm hair, purpose of, **2**:146;
      6:275–276
Underwear, labels on, **4**:4–5
Uniforms
   painters' whites, **6**:17–19
   surgeons', color of, **2**:86; **6**:269
United States Mint and shipment of
      coin sets, **5**:32
United States Postal System (USPS)
   CAR-RT SORT on envelopes,
      5:78–79
   undeliverable mail, **5**:13–14
University of Miami
   football helmets, **8**:171–172
   "Hurricanes" nickname, **8**:171–172
University of North Carolina and Tar
      Heels, **8**:76–77
University of South Florida, location
      of, **4**:7–8
Unleaded gasoline, cost of, **1**:121–122
Unlisted telephone numbers, charges
      for, **9**:45–47
Unscented hair spray, smell of, **2**:184

MASTER INDEX OF IMPONDERABILITY

"Upper crust," **11**:13
Upper lips, groove on, **6**:42–43
UPS
    shipment of coin sets, **5**:32
    used trucks, nonexistence of, **4**:73–74
Urinals, ice in, **10**:232–234
Urination, running water and, **8**:183; **9**:229–234
Uvula, purpose of, **3**:129

Vaccination marks, hair growth and, **2**:186
Valve stems on fire hydrants, shape of, **4**:142–143
VCRs, counter numbers on, **4**:148–149
Vegetable oil, vegetables in, **6**:266
Vegetable oils, vegetables in, **4**:186
Vending machines
    bill counting and, **6**:271–272
    freezing of soft drinks in, **9**:10–11
    half dollars and, **3**:56–57
    pennies and, **3**:54–56
    placement of snacks in, **9**:162–164
    "Use Correct Change" light on, **9**:186–188
Vent windows, side, in automobiles, **6**:13–14
Vestments
    color of Catholic priests', **10**:77–79
    color of popes', **10**:79–80
Videocassette boxes, configuration of, **9**:35–36
Videocassette recorders
    counters on, **6**:272–273; **7**:243–244
    power surges and, **7**:244–245
Videotape recorders
    "play" and "record" switches on, **5**:23–24
    storms and, **5**:180–181
Videotape versus audiotape, two sides of, **3**:136–137

Videotapes, rental, two-tone signals on, **5**:144–145
Violin bows, white dots on frogs of, **4**:164–165; **9**:291
Virgin acrylic, **7**:97–98
Virgin olive oil, **3**:174–175
Vision, 20–20, **3**:143
Vitamins, measurement of, in foods, **6**:148–150
Voices
    causes of high and low, **2**:70
    elderly versus younger, **6**:24–25
    perception of, own versus others, **1**:95–96
Volkswagen Beetles, elimination of, **2**:192–194
Vomiting and horses, **6**:111–112

"Waffling," **11**:183
Wagon wheels in film, movement of, **2**:183
Waiters' tips and credit cards, **7**:133–135
Walking the plank, pirates and, **9**:37–42; **10**:273–274
Walking, race, judging of, **9**:20–23
Wall Street Journal, lack of photographs in, **4**:41–42
Warmth and its effect on pain, **3**:134–135
Warning labels, mattress tag, **2**:1–2
Warts, frogs and toads and, **10**:121–123
Washing machine agitators, movement of, **4**:56
Washing machines, top- versus bottom-loading, and detergent, **1**:159–165
Washington D.C., "J" Street in, **2**:71
Watch, versus clock, distinctions between, **4**:77–78
"Watch," origins of term, **4**:77
Watches and placement on left hand, **4**:134–135; **6**:271

Water
  bottled, expiration dates on, **9**:77–78
  chemical manufacture of, **5**:107–108
  clouds in tap water, **9**:126–127
  color of, **2**:213
Water faucets
  bathroom versus kitchen, **5**:244
  location of, hot versus cold, **4**:191–192
Water temperature
  effect on stain, **6**:77–78
  versus air temperature, perception of, **4**:184
Water towers
  height of, **5**:91–93
  winter and, **6**:38–40
Water, boiling, boiling, during home births, **6**:114–115; **7**:247–248; **8**:254
Water, cold, kitchen versus bathroom, **4**:151–152
Watermelon seeds, white versus black, **5**:94
Wax, whereabouts in dripless candles, **4**:182–183
"Wear his heart on his sleeve," **11**:128
"Weasel words," **11**:90–91
Weather [see also particular conditions]
  clear days following storms, **6**:125
  forecasting of, in different regions, **6**:267–268
  partly cloudy versus partly sunny, **1**:21–22
  smell of impending rain, **6**:170–171
  West coast versus East coast, **4**:174–175
Wedding etiquette, congratulations to bride and groom and, **4**:86
Wedding invitations, tissue paper in, **4**:116–117

Weigh stations, highway, predictable closure of, **4**:193–194
Wells, roundness of, **10**:228–229
Wendy's hamburgers, square shape of, **1**:113–115
Western Union telegrams
  exclamation marks and, **3**:76–77
  periods, **3**:77–78
Wet noses, dogs and, **4**:70–73
Wetness, effect of, on color, **4**:139
Whales, sperm, head oil in, **6**:87–89
Whiplash, delayed reaction of, **10**:128–130
Whips, cracking sound of, **2**:74
Whistling at sporting events, **3**:199
White Castle hamburgers, holes in, **1**:80–83
White chocolate, versus brown chocolate, **5**:134–135
"White elephant," **11**:191–192
White paint on homes, **2**:100–102
White pepper, source of, **2**:135–136
White wine, black grapes and, **1**:201–202
White-wall tires
  bluish tinge on, **6**:192–193
  thickness of, **2**:149
Wigwams near highway, **10**:216–218
Wind on lakes, effect of different times on, **4**:156–157
Window cleaning, newspapers and, **10**:33–36
Window envelopes, use by mass mailers, **2**:111
Windows, rear, of automobiles, **5**:143–144
Windshield wipers, buses versus automobile, **7**:28
Wine
  chianti and straw-covered bottles, **8**:33–35
  dryness of, **5**:141–142

temperature of serving, red versus white, **4**:95–97

white, black grapes and, **1**:201–202

Wine glasses, types of, **7**:123–125

Wine tasters, drunkenness in, **1**:239–241

Wing-tip shoes, holes in, **8**:44

Winter, first day of, **3**:139–141

Wisdom teeth, purpose of, **2**:137

Women

  dancing ability, **6**:218; **7**:199–202; **8**:239–240

  fainting, **10**:219–222

  feelings of coldness, **6**:218; **7**:198–199; **8**:238

  group restroom visits, **6**:217; **7**:183–192; **8**:237–238; **9**:277–278

  hair length of aging, **8**:192–197

  leg kicking when kissing, **6**:218; **7**:196–197; **9**:278; 299–300

  remote control usage, **6**:217; **7**:193–196

  spitting, **8**:226–227

  uncomfortable shoes, **1**:62–64

Wood, popping noise of, in fires, **4**:10

Woodpeckers, headaches and, **4**:44–45

Wool

  shrinkage of, versus wool, **6**:166–167

  smell of, when wet, **4**:158–159

Worms

  appearance on sidewalk after rain, **4**:109

  as fish food, **3**:110–112

  birds and, **10**:65–72

  in tequila bottles, **7**:88–89

  larvae and, **6**:269–270

  survival during winter, **4**:108

Wrapping

  Burger King sandwiches, **8**:112–113

  chewing gum, **8**:111–112

  gift box chocolate, **8**:122–123

Wrinkles on extremities, **2**:112

Wrists as perfume target, **6**:90

Wristwatches, placement on left hand, **4**:134–135; **6**:271

X

  as symbol for kiss, **1**:128–129

  as symbol in algebra, **9**:131–132

"X ray," **11**:49–50

"Xmas," origins of term, **2**:75

X-rated movies, versus XXX-rated movies, **2**:141–142

"XXX [liquor]," **11**:58

Yawning, contagiousness of, **2**:238; **3**:213–217; **8**:231–232

Yeast in bread, effect of, **3**:144–145

Yellow, aversion of insects to, **8**:158–159

Yellow Freight Systems, orange trucks of, **6**:64–65

Yellow lights, timing of, in traffic lights, **1**:109–112

Yellow Pages, advertisements in, **3**:60–63

Yellowing of fingernails, nail polish and, **7**:129–130

YKK on zippers, **5**:180

Yogurt

  fruit on bottom, **7**:83

  liquid atop, **7**:82

Zebras and riding by humans, **8**:139–141

ZIP code, addresses on envelopes and, **3**:44

"Zipper," **11**:129

Zippers, YKK on, **5**:180

Zodiac, different dates for signs in, **4**:27–28

Zoo animals, rocking in, **10**:279

Help!

We hate to end the book on a downbeat note, but we have to admit one dread fact: Imponderability is not yet smitten. Let's stamp it out.

"How?" you ask. Send us letters with your Imponderables, answers to Frustables, gushes of praise, and even your condemnations and corrections.

Join your inspired comrades and become a part of the wonderful world of *Imponderables*. If you are the first person to submit an Imponderable we use in the next volume, we'll send you a complimentary copy, along with an acknowledgment in the book.

Although we accept "snail mail," we strongly encourage you to e-mail us if possible. Because of the volume of mail, we can't always provide a personal response to every letter, but we'll try—a self-addressed stamped envelope doesn't hurt. We're much better with answering e-mail, although we fall far behind sometimes when work beckons.

Come visit us online at the *Imponderables* website, where you can pose Imponderables, read our blog, and find out what's happening at Imponderables Central. Send your correspondence, along with your name, address, and (optional) phone number to:

feldman@imponderables.com
http://www.imponderables.com

Imponderables
P.O. Box 116
Planetarium Station
New York, NY 10024-0116

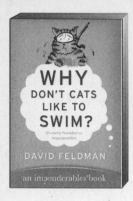 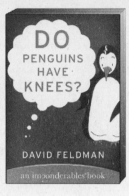